Speech Acoustic Analysis

Spoken Language Linguistics Set

coordinated by
Philippe Martin

Volume 1

Speech Acoustic Analysis

Philippe Martin

iSTE

WILEY

First published 2021 in Great Britain and the United States by ISTE Ltd and John Wiley & Sons, Inc.

ISTE Ltd
27-37 St George's Road
London SW19 4EU
UK

www.iste.co.uk

John Wiley & Sons, Inc.
111 River Street
Hoboken, NJ 07030
USA

www.wiley.com

Library of Congress Control Number: 2020948552

British Library Cataloguing-in-Publication Data
A CIP record for this book is available from the British Library
ISBN 978-1-78630-319-6

Contents

Preface

Courses in acoustic phonetics at universities are primarily aimed at humanities students, who are often skeptical about the need to assimilate the principles of physics that underlie the methods of analysis in phonetics. To avoid frightening anyone, many textbooks on phonetics are careful not to go into too much detail about the limitations and inner workings of the electronic circuits or algorithms used.

In writing this book, I have tried to maintain a different point of view. I am convinced that the basic principles of acoustic speech analysis can be easily explained without necessarily having to acquire a mathematical background, which would only be of interest to engineers studying telecommunications. To me, it seems more important to understand the analysis processes without necessarily knowing how to formalize them, in order to be able to fully master their properties and limitations and to face the practical problems of their use.

In fact, in terms of a mathematical background, it is usually enough to remember the notions learned in high school regarding elementary trigonometric functions, sine, cosine, tangent and cotangent, as well as logarithms (a reminder of their definitions can be found in the appendix). I hope that, with a reasonable effort in comprehension, readers will avoid errors and misinterpretations in the implementation and interpretation of acoustic measures, errors that are too often (and too late) found today in thesis defenses involving phonetic measures.

Illustrations involving the trigonometric functions were made using Graph software. Those presenting acoustic analysis results were obtained using WinPitch software. These various types of software can be downloaded for free from the Internet[1].

All that remains, is for me to thank the students of experimental phonetics in Aix-en-Provence, Toronto and Paris, as well as the excellent colleagues of these universities, for having allowed me, through their support, their criticisms and their suggestions, to progressively implement and improve the courses in acoustic phonetics that form the basis of this book. However, I am not forgetting G.B. (whose voice served as a basis for many examples), nor the merry band of doctoral students from Paris Diderot.

Philippe MARTIN
October 2020

1 www.padowan.dk and www.winpitch.com.

1

Sound

1.1. Acoustic phonetics

Phonetics is the science that aims to describe speech, phonology is the science that aims to describe language. Phonetics studies the sounds of human language from all angles, whereas phonology is only interested in these same sounds in terms of the role they have in the functioning of a language. Consequently, the objects described by phonetics are *a priori* independent of their function in the linguistic system, whether they are articulatory, acoustic or perceptive phonetics. While articulatory phonetics is very old (see the well-known scene from *Le Bourgeois Gentilhomme*, in which Monsieur Jourdain is having the details of the articulation of consonants and vowels explained to him very specifically, which every speaker realizes without being aware of the mechanisms involved), acoustic phonetics could only develop with the appearance of the first speech recording instruments, and the realization of instruments based on mathematical tools, to describe their physical properties.

During the 20th Century, recording techniques on vinyl disc, and then on magnetic tape, made it possible to preserve sound and analyze it, even in the absence of the speakers. Thanks to the development of electronics and the invention of the spectrograph, it was possible to quickly carry out harmonic analysis, sometimes painstakingly done by hand. Later, the emergence of personal computers in the 1980s, with faster and faster processors and large memory capacities, led to the development of computerized acoustic analysis tools that were made available to everyone, to the point that phonologists who were reluctant to investigate phonetics eventually used them. Acoustic phonetics aims to describe speech from a physical point of view, by explaining the characteristics that are likely to account for its use in the linguistic system. It also aims to describe the links between speech sounds and the phonatory mechanism, thus bridging the gap with traditional articulatory phonetics. Lastly, in the prosodic field, it is an essential tool for data acquisition, which is difficult to obtain reliably by auditory investigation alone.

1.2. Sound waves

Speech is a great human invention, because it allows us to communicate through sound, without necessarily having visual contact between the actors of communication. In acoustic phonetics, the object of which is the sound of speech, the term "sound" implies any perception by the ear (or both ears) of pressure variations in an environment in which these ears are immersed, in other words, in the air, but also occasionally under water for scuba divers. Pressure variations are created by a sound source, constituted by any material element that is in contact with the environment and manages to locally modify the pressure. In a vacuum, there can be no pressure or pressure variation. Sound does not propagate in a vacuum, therefore, a vacuum is an excellent sound insulator.

Pressure variation propagates *a priori* in all directions around the source, at a speed which depends on the nature of the environment, its temperature, average pressure, etc. In air that is 15°C, the propagation speed is 340 m/s (1,224 km/h) at sea level, while in seawater it is 1,500 m/s (5,400 km/h). Table 1.1 gives some values of propagation velocities in different environments. We can see that in steel, the speed of sound is one of the highest (5,200 m/s or 18,720 km/h), which explains why, in the movies from our childhood, outlaws of the American West could put an ear to a train rail to estimate its approach before attacking it, without too much danger. The possibility of sound being perceived by humans depends on its frequency and intensity. If the frequency is too low, less than about 20 Hz (in other words, 20 cycles of vibration per second), the sound will not be perceived (this is called infrasound). If it is too high (above 16,000 Hz, however this value depends on the age of the ear), the sound will not be perceived either (it is then called ultrasound). Many mammals such as dogs, bats and dolphins do not have the same frequency perception ranges as humans, and can hear ultrasound up to 100,000 Hz. This value also depends on age. Recently, the use of high-frequency sound generators, that are very unpleasant for teenagers, was banned to prevent them from gathering in certain public places, whereas these sounds did not cause any discomfort to adults who cannot perceive them.

Materials	Speed of sound (in m/s)
Air	343
Water	1,480
Ice	3,200
Glass	5,300
Steel	5,200
Lead	1,200
Titanium	4,950
PVC (soft)	80
PVC (hard)	1,700
Concrete	3,100
Beech	3,300
Granite	6,200
Peridotite	7,700
Dry sand	10 to 300

Table 1.1. *Some examples of sound propagation speed in different materials at a temperature of 20°C and under pressure of an atmosphere*[1]

1.3. In search of pure sound

In 1790, in France, the French National Constituent Assembly envisaged the creation of a stable, simple and universal measurement system. Thus, the meter, taking up a universal length first defined by the Englishman John Wilkins in 1668, and then taken up by the Italian Burattini in 1675, was redefined in 1793 as the ten millionth part of a half-meridian (a meridian is

1 http://tpe-son-jvc.e-monsite.com/pages/propagation-du-son/vitesse-du-son.html.

an imaginary, large, half circle drawn on the globe connecting the poles, with the circumference of the earth reaching 40,000 km). At the same time, the gram was chosen as the weight of one cubic centimeter of pure water at zero degrees. The multiples and sub-multiples of these basic units, meter and gram, will always be obtained by multiplying or dividing by 10 (the number of our fingers, etc.). As for the unit of measurement of time, the second, only its sub-multiples will be defined by dividing by 10 (the millisecond for one thousandth of a second, the microsecond for one millionth of a second, etc.), while for the multiples of minutes, hours and days, multiples of 60 and 24 remain unchanged. The other units of physical quantities are derived from the basic units of the meter, gram (or kilogram), second and added later, the Ampere, a unit of electrical current, and the Kelvin, a unit of temperature. Thus, the unit of speed is the meter per second m/s and the unit of power, the Watt, is defined by the formula $kg\ m^2/s^3$, both derived from the basic units of the kilogram, meter and second.

However, it was necessary to define a unit of sound, the unit of frequency derived from the unit of time, the cycle per second, applying to any type of vibration and not necessarily to sound. In the 18th Century, musicians were the main "sound producers" (outside of speech) and it seemed natural to turn to them to define a unit. Since the musicians' reference is the musical note "A" (more precisely "A3", belonging to the 3rd octave), and since this "A3" is produced by a tuning fork used to tune instruments, it remained to physically describe this musical reference sound and give it the title of "pure sound", which would be a reference to its nature (its timbre) and its frequency. When, during the first half of the 19th Century, the sound vibrations produced by the tuning fork could be visualized (with the invention of the kymograph and the phonautograph), it was noted that their form strongly resembled a well-known mathematical function, the sinusoid. Adopting the sinusoid as a mathematical model describing a pure sound, whose general equation, if we make the sinusoidal vibration time-dependent, is $f(t) = A \sin(\omega t)$, all that remained was to specify the meaning of the parameters A and ω (ω is the Greek letter "omega").

Instead of adopting the "A3" of musicians, the definition which was still fluctuating at the time (in the range of 420 to 440 vibrations per second, see Table 1.2), we naturally used the unit of time, the second, to define the unit of pure sound: one sinusoidal vibration per second, corresponding to one cycle of sinusoidal pressure variation per second.

Year	Frequency (Hz)	Location
1495	506	Organ of Halberstadt Cathedral
1511	377	Schlick, organist in Heidelberg
1543	481	Saint Catherine, Hamburg
1601	395	Paris, Saint Gervais
1621	395	Soissons, Cathedral
1623	450	Sevenoaks, Knole House
1627	392	Meaux, Cathedral
1636	504	Mersenne, chapel tone
1636	563	Mersenne, room tone
1640	458	Franciscan organs in Vienna
1648	403	Mersenne *épinette*
1666	448	Gloucester Cathedral
1680	450	Canterbury Cathedral
1682	408	Tarbes, Cathedral
1688	489	Hamburg, Saint Jacques
1690	442	London, Hampton Court Palace
1698	445	Cambridge, University Church
1708	443	Cambridge, Trinity College
1711	407	Lille, Saint Maurice
1730	448	London, Westminster Abbey
1750	390	Dallery Organ of Valloires Abbey
1751	423	Handel's tuning fork
1780	422	Mozart's tuning fork
1810	423	Paris, medium tuning fork
1823	428	Comic Opera Paris
1834	440	Scheibler Stuttgart Congress
1856	449	Paris Opera Berlioz
1857	445	Naples, San Carlos
1859	435	French tuning fork, ministerial decree
1859	456	Vienna

Year	Frequency (Hz)	Location
1863	440	Tonempfindungen Helmholtz
1879	457	Steinway Pianos USA
1884	432	Italy (Verdi)
1885	435	Vienna Conference
1899	440	Covent Garden
1939	440	Normal international tuning fork
1953	440	London Conference
1975	440	Standard ISO 16:1975

Table 1.2. *Change in the frequency of the reference "A" over the centuries (source: data from (Haynes 2002))*

1.4. Amplitude, frequency, duration and phase

1.4.1. *Amplitude*

A pure sound is therefore described mathematically by a sinusoidal function: sin(θ), θ (θ = Greek letter "theta") being the argument angle of the sine, varying as a function of time by posing θ = ωt (see Figure 1.1). To characterize the amplitude of the sound vibration, we multiply the sine function by a coefficient A (A for Amplitude): A sin(θ). Therefore, the greater the parameter A, the greater the vibration.

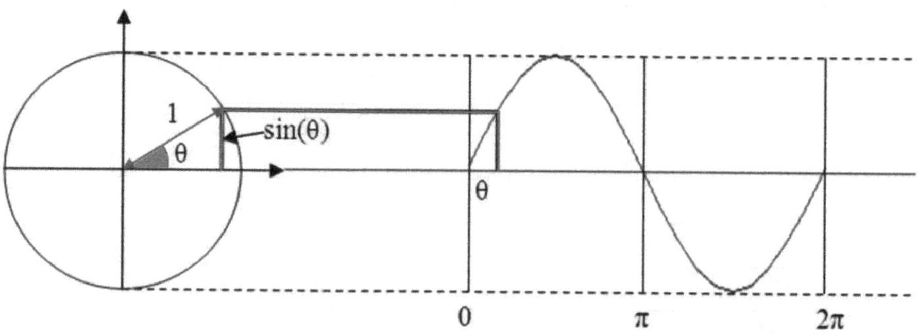

Figure 1.1. *Definition of the sinusoid. For a color version of this figure, see www.iste.co.uk/martin/speech.zip*

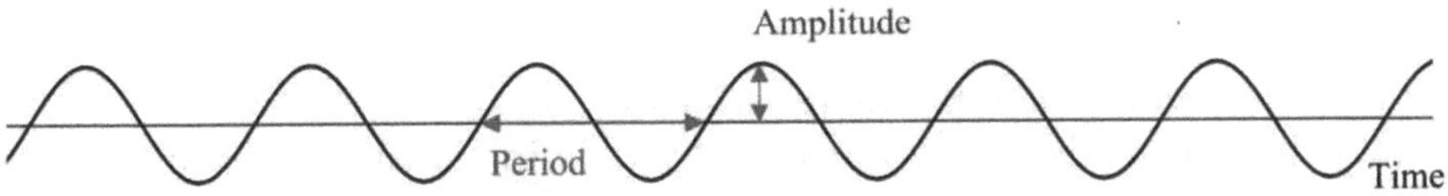

Figure 1.2. *Representation of pure sound as a function of time. For a color version of this figure, see www.iste.co.uk/martin/speech.zip*

1.4.2. *Frequency*

A single vibration of a pure sound carried out in one second, therefore one complete cycle of the sinusoid per second, corresponds to one complete revolution in the trigonometric circle that defines the sinusoid, in other words, an angle of 360 degrees, or if we use the radian unit (preferred by mathematicians), 2π radian, therefore 2 times π = 3.14159... = 6.28318... (π being the Greek letter "pi"). This value refers to the length of the trigonometric circle with a radius equal to 1.

A pure sound of one vibration per second will then have a mathematical representation of A $\sin(2\pi t)$, for if t = 1 second, the formula becomes A $\sin(2\pi)$. If the pure sound has two vibrations per second, the sinusoidal variations will be twice as fast, and the angle that defines the sinus will vary twice as fast in the trigonometric circle: A $\sin(2*2\pi t)$. If the sinusoidal variation is 10 times per second, the formula becomes A $\sin(10*2\pi t)$. This speed of variation is obviously called frequency and is represented by the symbol f (for "frequency"): A $\sin(2\pi ft)$.

By definition, a periodic event such as a pure sound is reproduced identically in time after a duration, called the period (symbol T, T for "time"). Frequency f and period T are opposite to one another: a pure (thus periodic) sound whose cycle is reproduced 10 times per second has a period 10 times smaller than one second, in other words, one tenth of a second, or 0.1 second, or 100 thousandths of a second (100 milliseconds or in scientific notation, 100 ms). If the period of a pure sound is 5 seconds, its frequency is one fifth of a cycle per second. The formula linking frequency and period is therefore $f = 1/T$, and therefore, also $T = 1/f$: the frequency is equal to 1 divided by the period, and conversely, the period is equal to 1 divided by the frequency.

1.4.3. *Duration*

Figure 1.2 could be misleading, in that it seems to limit the duration of a pure sound. In reality, pure sound is, by definition, infinite in duration, in other words, it begins in time − infinity and continues until + infinity of time. If this duration is truncated, for example limited to four periods, it is no longer a pure sound; and we will have to take this into account when manipulating its definition. This is one of the problems of using pure sound as a reference, since this object does not correspond to anything in the real sound world. One could have chosen another type of sound vibration that could be declared a "reference sound", but the tradition remains unwavering for the time being.

1.4.4. *Phase*

Pure sound, as defined by a sinusoidal function, is a mathematical idealization describing the evolution of an event over time, an event whose origin has not been determined, which can only be arbitrary.

The shift between this arbitrary origin and the starting point of a sinusoid reproduced in each cycle of pure sound, constitutes the phase (symbol φ, the Greek letter "phi" for phase). We can also consider the differences in the starting points of the time cycles of the different pure sounds. These differences are called phase shifts.

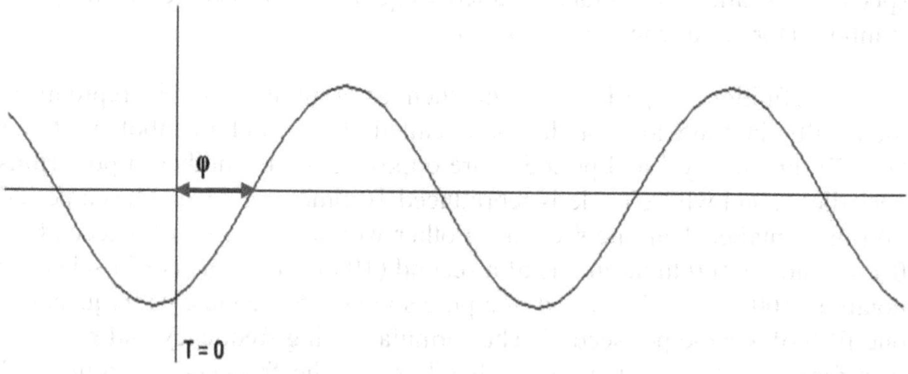

Figure 1.3. *Phase of pure sound*

A single pure sound will therefore only have a phase in relation to a temporal reference, and will be expressed in angles or time. The phase φ corresponds to a time shift Δt, which is related to the frequency f according to the formula $\Delta t = \Delta\varphi T/f = \Delta\varphi T$ (since $1/f = T$), with the Greek letter Δ "delta" for "difference".

When describing several pure tones of different amplitudes and frequencies, the phase parameter will be used to characterize the time shift of these pure tones with respect to each other.

The general mathematical representation of pure sound is enriched by the symbol φ, which is added to the argument of the sinusoid: $A \sin(2\pi ft + \varphi)$ if the frequency parameter is made explicit, and $A \sin((2\pi t/T) + \varphi)$ if the period is made explicit.

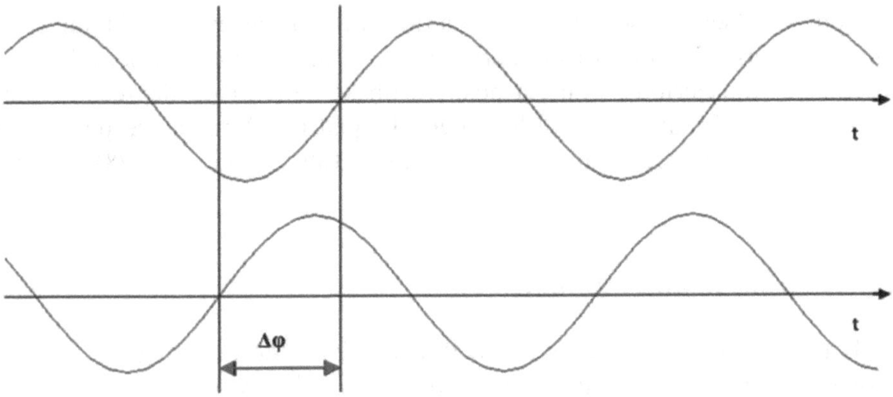

Figure 1.4. *Time shift due to phase shift*

1.5. Units of pure sound

Pure sound, a purely mathematical concept chosen to represent a reference of sound, is therefore characterized by three parameters: the amplitude of vibration, symbol A, the frequency of vibration, symbol f, and the phase φ of the vibration, describing the offset of the vibration with respect to an arbitrary reference point of time.

The unit of period is derived from the unit of time, the second. In practice, sub-multiples of the second, the millisecond or thousandth of a second (symbol ms), are used, particularly in acoustic phonetics. This sub-multiple adequately corresponds to quasi-periodic events related to the production of speech, such as the vibration of the vocal folds (in other words, the “vocal chords”), which typically open and close 70 to 300 times per second (sometimes faster when singing). In the early days of instrumental phonetics, the hundredth of a second was used as a unit instead (symbol cs). At that time, laryngeal cycle durations were in the order of 0.3 to 1.5 cs, which are now noted as 3 to 15 ms.

For frequency, the opposite of period, phoneticians have long since used cycles per second (cps) as the unit, but the development of the physics of periodic events eventually imposed the Hertz (symbol Hz).

For the phase, as the offset to a reference temporal origin must be specified, the units of angles (degree, grade or radian) are perfectly suitable. The phase offset can be converted to a time value if necessary, obtaining the time offset as a fraction of the period (or as a multiple of the period plus a fraction of the period). Thus, a positive phase shift of 45 degrees of pure sound at 100 Hz, in other words, a period equal to 10 ms, corresponds to a time shift with respect to the reference of (45/360) × 10 ms = (1/8) × 10 ms = 0.125 × 10 ms = 1.25 ms.

1.6. Amplitude and intensity

If the unit of frequency derived directly from the unit of time is not a problem, then what about the amplitude? In other words, what does an amplitude unit value correspond to? The answer refers to what the sinusoidal equation of pure sound represents, in other words, a change in sound pressure. In physics, the unit of pressure is defined as a unit of force applied perpendicularly to a unit of area. In mechanics, and therefore also in acoustics, the unit of force is the Newton (in honor of Isaac Newton and his apple, 1643–1727), and one Newton (symbol N) is defined as equal to the force capable of delivering an increase in speed of 1 meter per second every second (thus an acceleration of 1 m/s^2) to a mass (an apple?) of 1 kilogram.

By comparing all these definitions, we obtain, for the unit of pressure, the Pascal (symbol Pa, in memory of Blaise Pascal, 1623–1662): $1\text{ Pa} = 1\ N/m^2$ or, by replacing the Newton by its definition in basic units MKS, Meter, Kilogram, Second, $1\text{ Pa} = 1\ kg/s^2m^3$. Compared to atmospheric pressure, which is on average 100,000 Pa (1,000 hecto Pascal or 1,000 hPa, hecto symbol h means 100 in Greek), sound pressures are very small, but vary over a wide range from about 20 µPa (10 micro Pascal, in other words, 20 millionths of a Pascal) to 20 Pa, that is to say, a factor of 1 to 1,000,000! At normal conversation levels, the maximum sound pressure variation reaching our ears is about 1 Pa.

So much for the amplitude of pressure variation of pure sound, which is expressed in Pascal. But what about the intensity? Physics teaches us that intensity is defined by the power of the vibration, divided by the surface on which it is applied. In the case of pure sound, this leads us to the calculation of the power delivered by a sinusoidal pressure variation, in other words, the amount of energy delivered (or received) per unit of time.

The unit of energy is the Joule (symbol J, in honor of the English physicist James Prescott Joule, 1818–1889), equal to the work of a force of one Newton, whose point of application moves one meter in the direction of the force, so 1 J = 1 N m = 1 kg m^2/s^2 since 1 Newton = 1 m/s^2.

To get closer to the intensity, we still have to define the unit of power, the Watt (symbol W, from the name of the English engineer who invented the steam engine James Watt, 1736–1809). One Watt corresponds to the power of one Joule spent during one second: 1 W = 1 J/s, in other words = 1 N m/s, or 1 kg m^2/s^3.

The pressure of a pure sound, expressed in Pascal, varies around the average pressure at the place of measurement (for example, the atmospheric pressure at the eardrum) during a period of the sound from +A Pa to −A Pa (from a positive amplitude +A to a negative amplitude −A). Knowing that this variation is sinusoidal for a pure sound, one can calculate the average energy expended during a complete cycle, that is to say, for a period of one second, $A/\sqrt{2}$ (formula resulting from the integration of the two half-periods of a sinusoid).

Since the power is equal to the pressure (in Pa) multiplied by the displacement of the vibration (thus the amplitude A) and divided by the time $W = N A/s$, and the intensity is equal to the power divided by the area $I = W/m^2$, we deduce, by substituting W par $N A/s$, that the sound intensity is proportional to the square of the amplitude: $I \div A^2/s$. This formula is very important in order to understand the difference between amplitude and the intensity of a sound.

1.7. Bels and decibels

While the range of the pressure variation of a pure sound is in the order of 20 μPa to 20 Pa, in other words, a ratio of 1 to 1,000,000, that of the intensity variation corresponds to the square of the amplitude variation, i.e. a ratio of 1 to 1,000,000,000,000, or approximately from $10^{-12}\,\mathrm{W/m^2}$ to $1\,\mathrm{W/m^2}$. Using a surface measurement better suited to the eardrum, the $\mathrm{cm^2}$, the range of variation is then expressed as $10^{-16}\,\mathrm{W/cm^2}$ to $10^{-4}\,\mathrm{W/cm^2}$. Developed at a time when mechanical calculating machines were struggling to provide all the necessary decimals (were they really necessary?), the preferred method was to use a conversion that would allow the use of less cumbersome values, and also (see below) one that reflected the characteristics of the human perception of pure sounds, to some extent. This conversion is the logarithm.

The most common logarithm (there are several kinds) used in acoustics, is the logarithm to base 10 (log10 or log notation), equal to the power to which the number 10 must be raised, to find the number whose logarithm is desired.

We therefore have:

1) log(1) = 0 since $10^0 = 1$ (10 to the power of zero equals 1);

2) log(10) = 1 since, $10^1 = 10$ (10 to the power of 1 equals 10);

3) log(100) = 2 since, $10^2 = 100$ (10 to the power of 2 equals 10 times 10, in other words, 100);

4) log(1000) = 3 since, $10^3 = 1{,}000$ (10 to the power 3 equals 10 times 10, in other words, 1,000).

For values smaller than 1:

1) log(0,1) = −1 since, $10^{-1} = 1/10$ (negative exponents correspond to 1 divided by the value with a positive exponent);

2) log(0.01) = −2 since, $10^{-2} = 1/100$.

The fact remains that the value of the logarithm of numbers other than the integer powers of 10 require a rough calculation. The calculation of log(2), for example, can be done without a calculator, by noting that $2^{10} = 1\ 024$, so 10 log(2) = approximately 3 (actually 3.01029...), so log(2) = approximately 0.3.

Another advantage of switching to logarithms is that the logarithm of the multiplication of two numbers is transformed into the addition of their logarithm: log(xy) = log(x) + log(y). This property is at the basis of the invention of slide rules, allowing the rapid multiplication and division of two numbers by sliding two rules graduated in logarithm. These rules, which made the heyday of the generation of engineers, are obviously abandoned today in favor of pocket-sized electronic calculators or smartphones.

1.8. Audibility threshold and pain threshold

The difference in intensity between the weakest and loudest sound that can be perceived, in other words, between the threshold of audibility and the so-called pain threshold (beyond which the hearing system can be irreversibly damaged), is therefore between 1 and 1,000,000,000,000. In order to use a logarithmic scale representing this range of variation, we need to define a reference, since the logarithm of an intensity has no direct physical meaning as it relates to a ratio of an intensity value, to an intensity taken as a reference. The first reference value that comes to mind is the audibility threshold (corresponding to the lowest intensity of sound that can be perceived), but chosen at a frequency of 1,000 Hz. It was not known in the 1930s that human hearing was even more sensitive than believed, in the region of 2,000 Hz to 5,000 Hz, and that this threshold was therefore even lower. It was arbitrarily decided that this threshold would have a reference value of 20 μPa, which is assigned the logarithmic value of 0 (since log (20 μPa /20 μPa) = log (1) = 0).

Since the researchers at the research centers of the American company Bell Telephone Laboratories, H. Fletcher and W.A. Munson, were heavily involved in research on the perception of pure sounds, the Bel was chosen as the unit (symbol B), giving the perception threshold at 1,000 Hz a 0 Bel value. Since the pain threshold is 1,000,000,000,000 stronger in intensity, its value in Bel is expressed by the ratio of this value to the reference of the perception threshold whose logarithm is calculated, in other words, log(1,000,000,000,000/1) = 12 B. Using the pressure ratio gives the same result (remember that the intensity is proportional to the square of the amplitude): log (20 Pa/20 μPa) = log (20,000,000 μPa/20 μPa) = log (1,000,000) = 6 B for the amplitude ratio, and 2 x 6 B = 12 B for the intensity ratio, since the logarithm of the square of the amplitude is equal to 2 times the logarithm of the amplitude. The Bel unit seems a little too large in practice, so we prefer to use tenths of Bel, or decibel, symbol dB. This time, the ratio of variation between the strongest and weakest sound in amplitude is 60 dB with an intensity of 120 dB.

A remarkable value to remember is the decibel increase resulting from doubling the amplitude of pure sound: 10 log (2) = 3 dB for amplitude and 20 log (2) = 6 dB for intensity. The halving of the amplitude causes a drop in amplitude of −3 dB and in intensity of −6 dB. Multiplying the amplitude by a factor of 10 corresponds to an increase in intensity of 20 log(10) = 20 dB, by 100 of 40 dB, etc.

The dB unit is always a relative value. To avoid any ambiguity, when the implicit reference is the hearing threshold, we speak of absolute decibels (dB SPL, with SPL standing for sound pressure level) or relative decibels. Absolute dBs are therefore relative dBs with respect to the audibility threshold at 1,000 Hz.

1.9. Intensity and distance from the sound source

The decrease in the intensity of a pure sound decreases with the square of the distance r. This is easily explained in the case of radial propagation of the sound in all directions around the source. If we neglect the energy losses during the propagation of sound in air, the total intensity all around the source is constant. Since the propagation is spherical, the surface of the sphere increases and is proportional to the square of its radius r, in other words, the distance to the source, according to the (well known) formula $4 \pi r^2$. The intensity of the source (in a lossless physical model) is therefore distributed over the entire surface and its decrease is proportional to the square of the distance from the sound source.

$I \div 1/r^2$. The amplitude of a pure sound therefore decreases inversely to the distance, since the intensity I is proportional to the square of the amplitude A. However, we have $A^2 \div 1/r^2$ and A $\div \sqrt{1/r^2}$ so A $\div 1/r$.

The relationship of the amplitude to the distance from the sound source is of great importance for sound recording. For example, doubling the distance between a speaker and the recording microphone means decreasing the amplitude by a factor of two and the intensity by a factor of four. While the optimal distance for speech recording is about 30 cm from the sound source, placing a microphone at a distance of 1 m results in a drop in amplitude by a factor of 3.33 (5.2 dB) and intensity by a factor of 10 (20 dB).

1.10. Pure sound and musical sound: the scale in Western music

In the tempered musical scale, the note frequencies are given by the following formula:

$$f = ref\ 2^{((octave-3)+(tone-10)/12)}$$

where *octave* and *tone* are integers, and *ref* is the reference frequency of 440 Hz. Table 1.3 gives the frequencies of the notes in the octave of the reference *A* (octave 3). The frequencies must be multiplied by two for an octave above, and divided by two for an octave below.

Notes	Frequency
B# / C	261.6 Hz
C# / D*b*	277.2 Hz
D	293.7 Hz
D# / E*b*	311.1 Hz
E / F*b*	329.7 Hz
E# / F	349.2 Hz
F# / G*b*	370.0 Hz
G	392.0 Hz
G# / A*b*	415.3 Hz
A	440.0 Hz
A# / B*b*	466.2 Hz
B / C*b*	493.9 Hz

Table 1.3. *Frequencies of musical notes*

1.11. Audiometry

Fletcher-Munson curves, developed in the 1930s from perceptual tests conducted on a relatively large population, give values of equal intensity perceived as a function of frequency. It was realized then, that the 1,000 Hz value used as a reference for defining dB may not be optimal, since the average sensitivity of the ear is better in the frequency range of 2,000 Hz to 5,000 Hz. Thus, we have hearing thresholds at 4,000 Hz, that are negative in dB (about −5 dB) and therefore lower than the 0 dB reference! Perceived curves of equal intensity involve delicate measurements (listeners should be asked to judge the equality of intensity of two pure sounds of different frequencies). They were revised in 1956 by Robinson and Dadson, and were adopted by the ISO 226: 2003 standard (Figure 1.5).

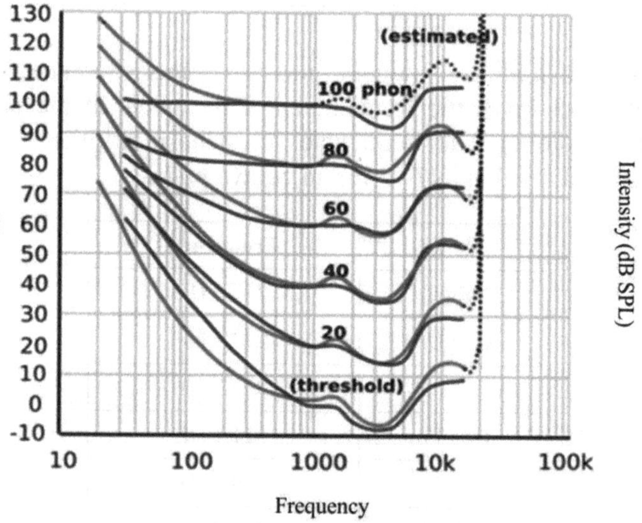

Figure 1.5. *Curves of equal intensity perceived versus Fletcher-Munson frequency (in blue) reviewed by Robinson and Dadson (in red) (source: (Robinson and Dadson 1956)). For a color version of this figure, see www.iste.co.uk/martin/speech.zip*

These curves show a new unit, the Phon, attached to each of the equal-perception curves, and corresponding to the values in dB SPL at 1,000 Hz. From the graph, we can see, for example, that it takes 10 times as much intensity at 100 Hz (20 dB), than at 1,000 Hz, to obtain the same sensation of intensity for pure sound at 40 dB SPL. We can also see that the zone of maximum sensitivity is between 2,000 Hz and 5,000 Hz, and that the pain threshold is much higher for low frequencies, which, on the other hand, allows a lower dynamic range (about 60 dB) compared to high frequencies (about 120 dB).

Another unit, loudness, was proposed by S. Smith Stevens in 1936, so that the doubling of the loudness value corresponds to the doubling of the perceived intensity. The correspondence between loudness and phons is made at 1,000 Hz and 40 phons, or 40 dB SPL, equivalent to 1 loudness. Table 1.4 gives other correspondence values. Pitch tones are rarely used in acoustic phonetics.

Loudness	1	2	4	8	16	32	64
Phons	40	50	60	70	80	90	100

Table 1.4. *Correspondence between loudness and phons*

1.12. Masking effect

Two pure sounds perceived simultaneously can mask each other, in other words, only one of them will be perceived. The masking effect depends on the difference in frequency and in intensity of the sounds involved. It can also be said that the masking effect modifies the audibility threshold locally, as shown in Figure 1.6.

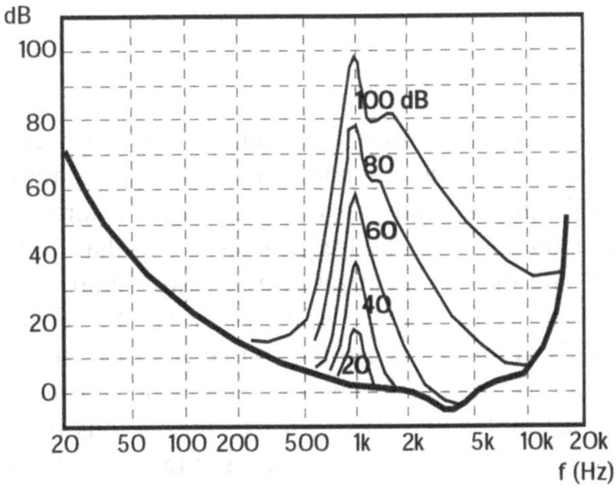

Figure 1.6. *Modification of the audibility threshold by the simultaneous masking effect according to a masking sound at 1,000 Hz (from (Haas 1972))*

There is also temporal masking, in which a sound is masked, either by another sound that precedes it (precedence masking, or Haas effect, according to Helmut Haas), or by another sound that follows it (posteriority masking). This type of masking only occurs for very short sounds, in the order of 50 to 100 ms (see Figure 1.7).

The masking effect, simultaneous and temporal, is used intensively in algorithms for speech and music compression (MP3, WMA, etc. standards). Oddly enough, few works in acoustic phonetics make explicit use of it, nor do Fletcher-Munson's equal-perception curves.

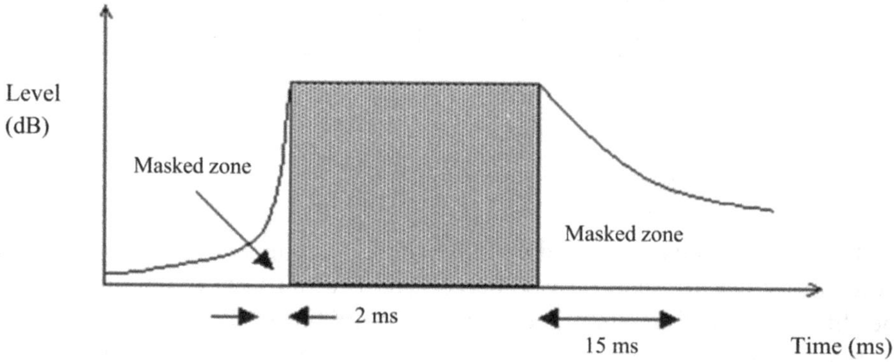

Figure 1.7. *Temporal mask effect (from (Haas 1972))*

1.13. Pure untraceable sound

As we have seen, the search for a unit of sound was based on the reference used by the musicians and generated by the tuning fork. Pure sound consists of a generalization towards an infinite duration of the sound produced by the tuning fork, at frequencies other than the reference A3 (today, 440 Hz), and thus an idealization that pure sound is infinite in time, both in the past and in the future. On the contrary, the sound of the tuning fork begins at a given instant, when the source is struck in such a way as to produce a vibration of the metal tube, a vibration which then propagates to the surrounding air molecules. Then, due to the various energy losses, the amplitude of the vibration slowly decreases and fades away completely after a relatively long (more than a minute), but certainly not infinite, period of time. This is referred to as damped vibration (Figure 1.8).

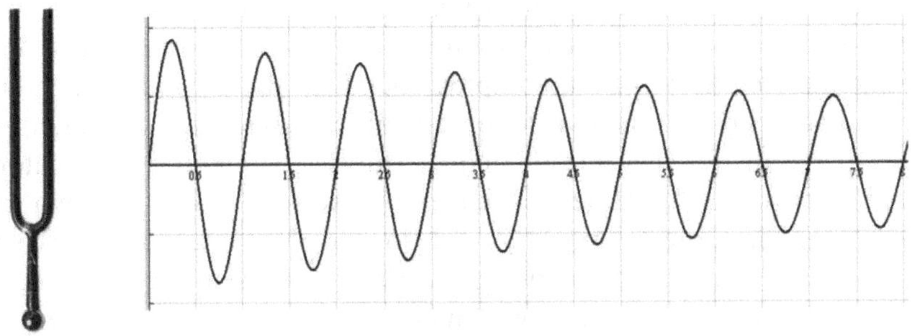

Tuning fork Damped vibration

Figure 1.8. *The tuning fork produces a damped sinusoidal sound variation*

From this, it will be remembered that pure sound does not actually exist, since it has no duration (or an infinite duration), and yet, perhaps under the weight of tradition, and despite the recurrent attempts of some acousticians, this mathematical construction continues to serve as the basis of a unit of sound for the description and acoustic measurement of real sounds, and in particular of speech sounds.

Apart from its infinite character (it has always existed, and will always exist... mathematically), because of its value of 1 Hz in frequency and because of its linear scale in Pascal for amplitude, pure sound does not really seem to be well suited to describe the sounds used in speech. Yet this is the definition that continues to be used today.

1.14. Pure sound, complex sound

In any case, for the moment, the physical unit of sound, pure sound, is a sinusoidal pressure variation with a frequency of 1 Hz, and an amplitude equal to 1 Pa (1 Pa = $\frac{1kg}{s^2m^3}$).

What happens when we add two pure sounds of different frequencies? There are two obvious cases: 1) either the frequency of one of the pure sounds is an integer multiple of the frequency of the first one, and we will then say that this pure sound is a harmonic of the first one (or that it has a harmonic frequency of the frequency of the first one), or 2) this frequency is not an integer multiple of the frequency of the first sound.

In the first case, the addition of the two pure sounds gives a “complex” sound, where the frequency of the first sound corresponds to the fundamental frequency of the complex sound. In the second case (although we can always say that the two sounds are always in a harmonic relationship, because it is always possible to find the lowest common denominator that corresponds to their fundamental frequency), we will say that the two sounds are not in a harmonic relationship and do not constitute a complex sound. Later, we will see that these two possibilities of frequency ratio between pure sounds characterize the two main methods of acoustic analysis of speech: Fourier’s analysis (Jean-Baptiste Joseph Fourier, 1768–1830), and Prony’s method, also called LPC (Gaspard François Clair Marie, Baron Riche de Prony, 1755–1839).

It is natural to generalize the two cases of pure sounds assembly to an infinity of pure sounds (after all, we live in the idealized world of physics models), whose frequencies are in a harmonic ratio (thus integer multiples of the fundamental frequency), or whose frequencies are not in a harmonic ratio.

In the harmonic case, this assembly is described by a mathematical formula using the Σ symbol of the sum:

$$\sum_{n=0}^{N} A_n \sin(n\omega t + \varphi_n)$$

with $\omega = 2\pi f$ (the pulse) and φ = the phase, in other words, a sum of N pure sounds of multiple harmonic frequencies of the parameter n, which varies in the formula from 0 to N, and are out of phase with each other. According to this formula, the fundamental has an amplitude A_1 (the amplitude of the first component sinusoid), a frequency $\omega/2\pi$ and a phase φ_1. The zero value of n corresponds to the so-called continuous component with amplitude A_0 and zero frequency. A harmonic of the order n has an amplitude A_n, a frequency $n\omega/2\pi$ and a phase φ_n. This sum of harmonic sounds is called a harmonic series or Fourier series. Figures 1.9 and 1.10 show examples of a 3-component harmonic series and the complex sound obtained by adding the components with different phases. A complex sound is therefore the sum of harmonic sounds, which are integer multiples of the fundamental frequency.

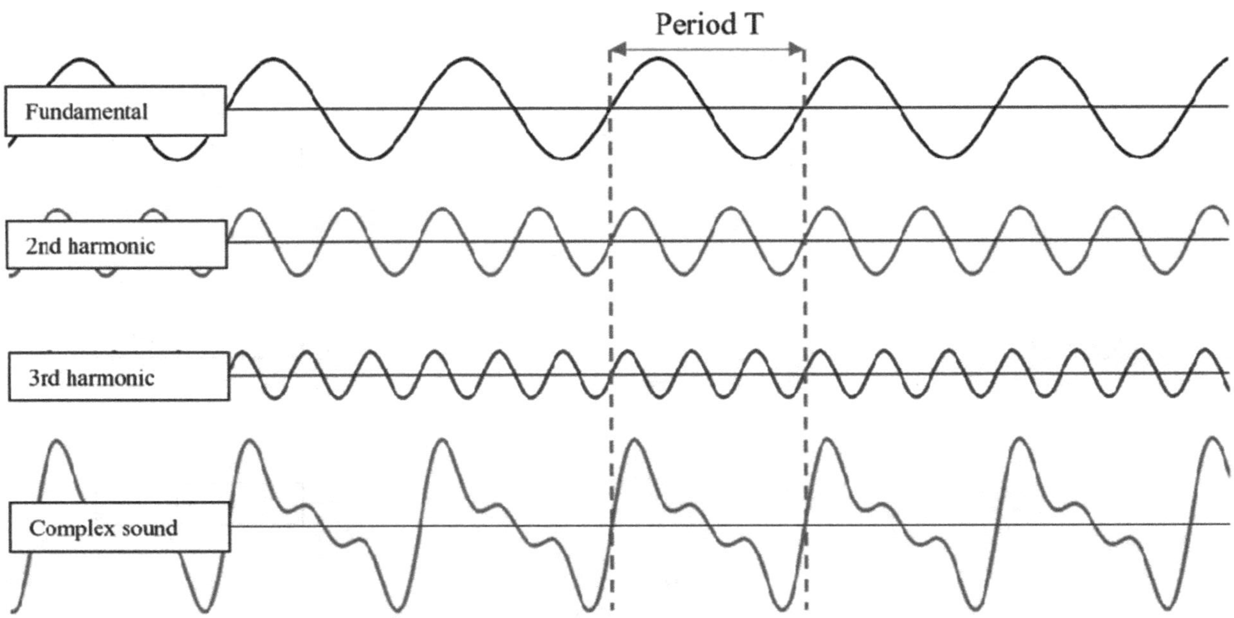

Figure 1.9. *Example of a complex sound constituted by the sum of 3 pure sounds of in-phase harmonic frequencies. For a color version of this figure, see www.iste.co.uk/martin/speech.zip*

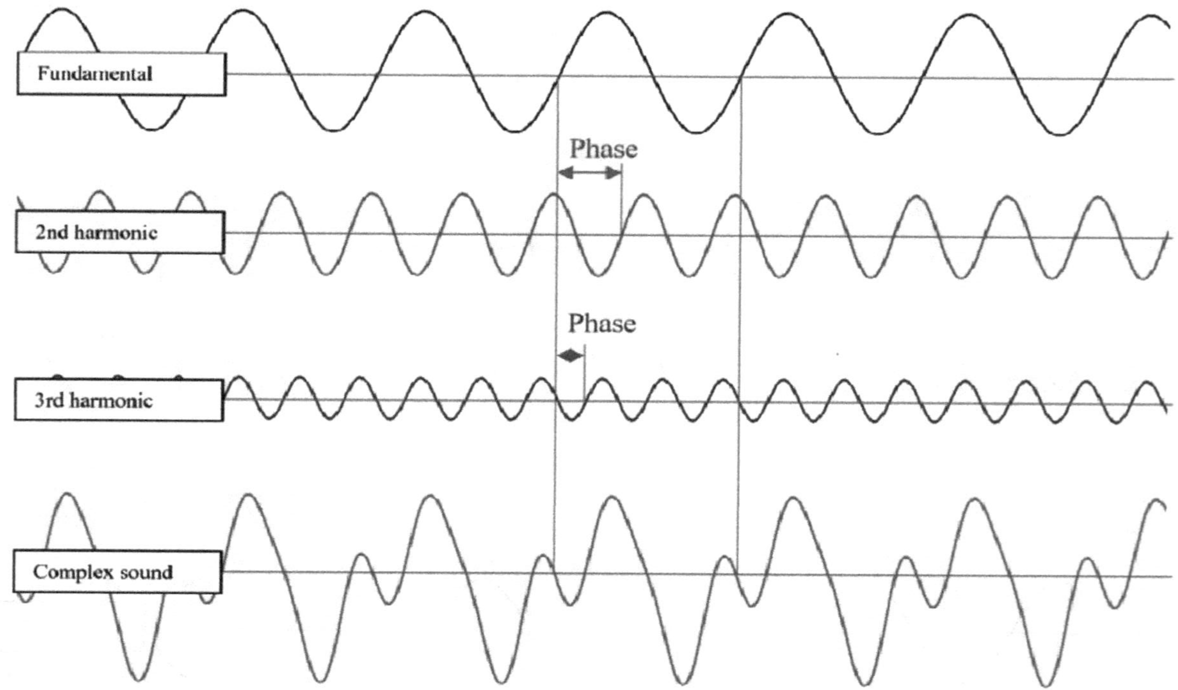

Figure 1.10. *Example of a complex sound constituted by the sum of 3 pure sounds of harmonic frequencies out of phase with the fundamental. For a color version of this figure, see www.iste.co.uk/martin/speech.zip*

For the Roman musicologist Estorc, "it was not until the beginning of the 11th Century A.D. that Gui d'Arezzo, in his work Micrologus, around 1026, developed the theory of solmization, with the names we know (do, re, mi, fa, sol, la, ti) and put forward the idea of an equal note at all times at the same pitch".

Thus, over time, the idea of creating a precise, immutable note to tune to emerged. But what frequency was to be chosen? It depended on the instruments, the nature of the materials used, and also on regionalism and times. Romain Estorc continues: "For 16th Century music, we use la 466 Hz, for Venetian Baroque (at the time of Vivaldi), it's la 440 Hz, for German Baroque (at the time of Telemann, Johann Sebastian Bach, etc.), it's la 415 Hz, for French Baroque (Couperin, Marais, Charpentier, etc.) we tune to la 392 Hz! There are different pitches such as Handel's tuning fork at 423 Hz, Mozart tuning fork at 422 Hz, that of the Paris Opera, known as Berlioz, at 449 Hz, that of Steinway pianos in the USA, at 457 Hz."

The beginnings of this rationalization appeared in 1884, as Amaury Cambuzat points out, "when the composer Giuseppe Verdi obtained a legal decree from the Italian government's musical commission, normalizing the pitch to a *la* (i.e. *A*) at 432 vibrations per second" This decree is exhibited at the Giuseppe-Verdi Conservatory in Milan. It was unanimously approved by the commission of Italian musicians.

Thanks to Verdi, the 432 Hz made its appearance as a reference at the end of the 19th Century. In 1939, there was a change of gear: the International Federation of National Standardization Associations, now known as the International Organization for Standardization, decided on a tuning fork standard-meter at 440 Hz.

This decision was approved a few years later, at an international conference in London in 1953, despite the protests of the Italians and the French, who were attached to Verdi's *la* 432 Hz. Finally, in January 1975, the *la* 440 Hz pitch became a standard (ISO 16:1975), which subsequently defined its use in all music conservatories. The 440 Hz frequency thus won the institutional battle, establishing itself as an international standard.

(From Ursula Michel (2016) "432 vs. 440 Hz, the astonishing history of the frequency war", published on September 17, 2016)[2].

Box 1.1. *The "la" war*

2 http://www.slate.fr/story/118605/frequences-musique.

2

Sound Conservation

2.1. Phonautograph

In addition to the analysis of speech by observing the physiological characteristics of the speaker, acoustic analysis has the great advantage of not being intrusive (at least from a physical point of view, whereas speech recording can be intrusive from a psychological point of view), and of allowing the data to be easily stored and further processed without requiring the presence of the speaking subject.

The recording of acoustic data is carried out by a series of processes, the first of which consists of transforming the air pressure variations that produce sound into variations of another nature, whether that be mechanical, electrical, magnetic or digital. These may be converted back into pressure variations by means of earphones, or a loudspeaker, in order to more or less faithfully reconstitute the original acoustic signal.

The first recording systems date back to the beginning of the 19th Century, a century that was one of considerable development in mechanics, while the 20th Century was that of electronics, and the 21st Century is proving to be that of computers. The very first (known) sound conservation system was devised by Thomas Young (1773–1829), but the most famous achievement of this period is that of Édouard–Léon Scott de Martinville (1817–1879) who, in 1853, through the use of an acoustic horn, succeeded in transforming the sound vibrations into the vibrations of a writing needle tracing a groove on a support that was moved according to time (paper covered in smoke black rolled up on a cylinder). Scott de Martinville called his device a phonautograph (Figure 2.1). The phonautograph received many refinements in the following years. Later, in 1878, Heinrich Schneebeli produced excellent vowel tracings that made it possible to perform a Fourier harmonic analysis for the first time (Teston 2006).

Figure 2.1. *Scott de Martinville's phonautograph, Teylers Museum, Haarlem, The Netherlands*[1]. *For a color version of this figure, see www.iste.co.uk/martin/speech.zip*

This method could not reproduce sounds, but in 2007 enthusiasts of old recordings were able to digitally reconstruct the vibrations recorded on paper by optical reading of the oscillations and thus allow them to be listened to (Figure 2.2)[2]. Later, it was Thomas Edison (1847–1931) who, in 1877, succeeded in making a recording without the use of paper, but instead on a cylinder covered with tin foil, which allowed the reverse operation of transforming the mechanical vibrations of a stylus travelling along the recorded groove into sound vibrations. Charles Cros (1842–1888) had filed a patent describing a similar device in 1877, but had not made it.

Later, the tin foil was replaced by wax, then by bakelite, which was much more resistant and allowed many sound reproductions to be made without

1 https://www.napoleon.org/en/history-of-the-two-empires/objects/edouard-leon-scott-de-martinvilles-phonautographe/.

2 www.firstsounds.org.

destroying the recording groove. In 1898, Valdemar Poulsen (1869–1942) used a magnetized piano string passing at high speed in front of an electromagnet whose magnetization varied depending on the sound vibration. Later, in 1935, the wire was replaced by a magnetic tape on a synthetic support (the tape recorder), giving rise to the recent systems with magnetic tape. This system has been virtually abandoned since the appearance of digital memories that can store a digital equivalent of the oscillations.

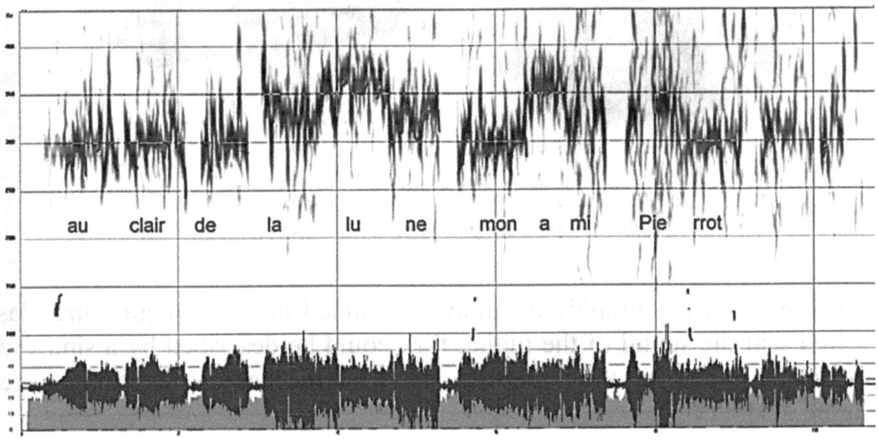

Figure 2.2. *Spectrogram of the first known Clair de la Lune recording, showing the harmonics and the evolution of the melodic curve (in red). For a color version of this figure, see www.iste.co.uk/martin/speech.zip*

2.2. Kymograph

The physiologists Carlo Matteucci, in 1846, and Carl Ludwig, in 1847, had already paved the way for sound recording. As its name suggests (from the Greek, κῦμα, swelling or wave, and γραφή, written), the kymograph is an instrument that was initially used to record temporal variations in blood pressure, and later extended to muscle movements and many other physiological phenomena. It consists of a rotary cylinder rotating at a constant speed as a function of time. Variations in the variable being studied, pressure variations, in the case of speech sounds, are translated into a linear variation of a stylus that leaves a trace on the paper covered in smoke black that winds around the cylinder. The novelty that Scott de Martinville brought to the phonautograph was the sound pressure sensor tube, which allowed a better trace of the variations due to the voice.

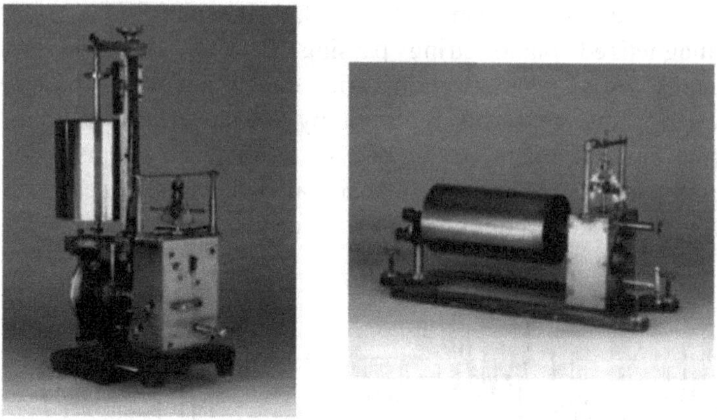

Figure 2.3. *Ludwig's kymographs (source: Ghasemzadeh and Zafari 2011). For a color version of this figure, see www.iste.co.uk/martin/speech.zip*

These first kymographic diagrams, examined under a magnifying glass, showed that the sound of the tuning fork could be described by a sinusoidal function, disregarding the imperfections of the mechanical recording system (Figure 2.4).

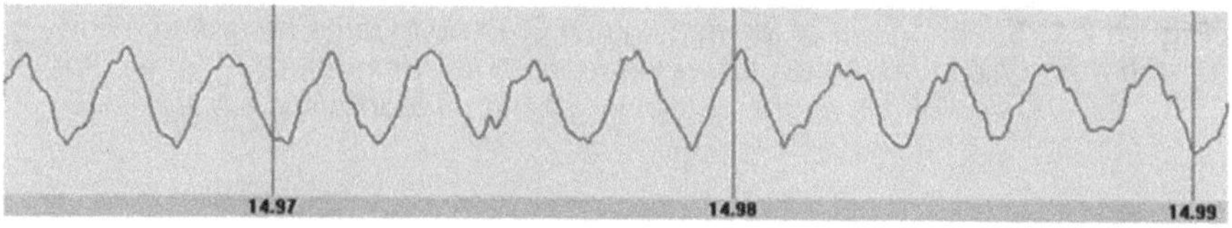

Figure 2.4. *Waveform and spectrogram of the first known recording of a tuning fork (by Scott de Martinville 1860). For a color version of this figure, see www.iste.co.uk/martin/speech.zip*

In his work, Scott de Martinville had also noticed that the complex waveform of vowels like [a] in Figure 2.5 could result from the addition of pure harmonic frequency sounds, paving the way for the spectral analysis of vowels (Figure 2.6).

The improvements of the kymograph multiplied and its use for the study of speech sounds was best known through the work of Abbé Rousselot (1846–1924), reported mainly in his work *Principes de phonétique expérimentale*, consisting of several volumes, published from 1897 to 1901 (Figure 2.7).

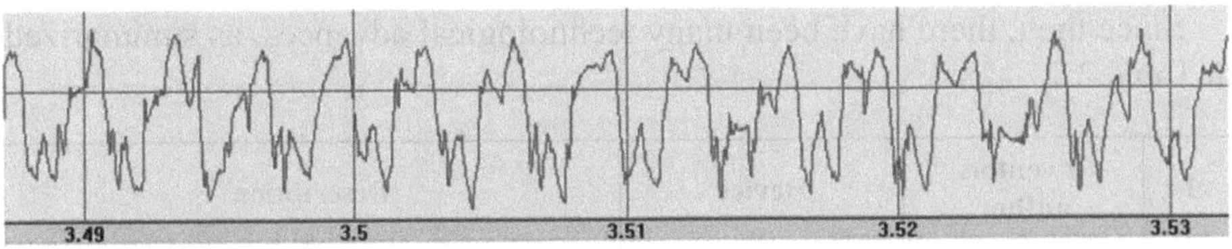

Figure 2.5. *Waveform of an [a] in the 1860 recording of Scott de Martinville. For a color version of this figure, see www.iste.co.uk/martin/speech.zip*

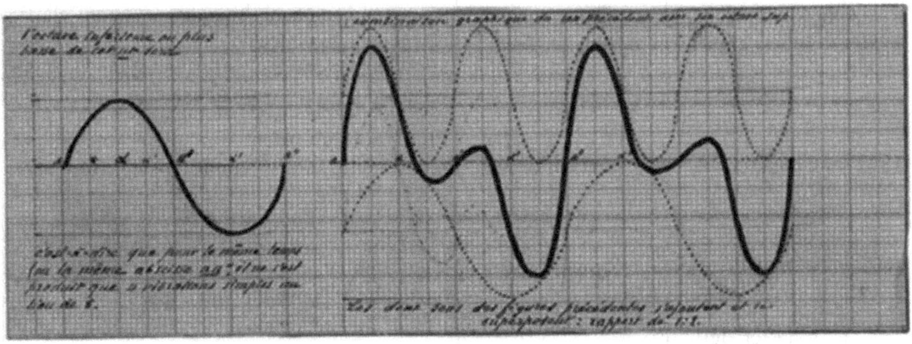

Figure 2.6. *Graphical waveform calculation resulting from the addition of pure tones by Scott de Martinville (© Académie des Sciences de l'Institut de France). For a color version of this figure, see www.iste.co.uk/martin/speech.zip*

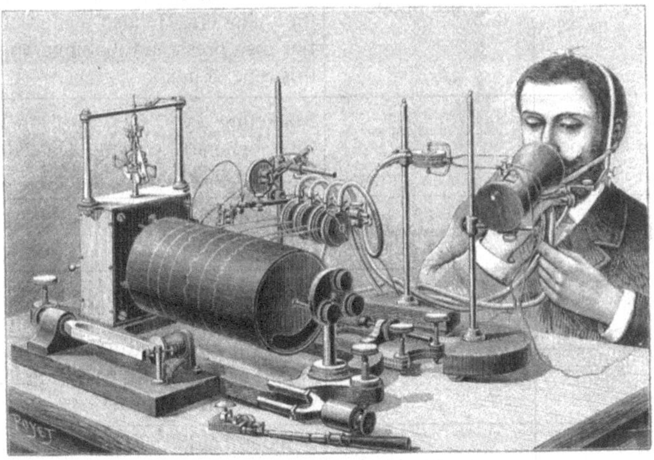

Appareil de M. l'abbé Rousselot pour l'inscription de la parole.

Figure 2.7. *La Nature, No. 998, July 16, 1892. Apparatus of M. l'abbé Rousselot for the inscription of speech*

Since then, there have been many technological advances, as summarized in Table 2.1.

1	Inventor, author	Device	Description
1807	Thomas Young		The Vibrograph traced the movement of a tuning fork against a revolving smoke-blackened cylinder.
1856	Léon Scott de Martinville	Phonautograph	Improving on Thomas Young's process, he managed to record the voice using an elastic membrane connected to the stylus and allowing engraving on a rotating cylinder wrapped with smoke-blackened paper.
1877	Charles Cros	Paleophone	Filing of a patent on the principle of reproducing sound vibration engraved on a steel cylinder.
1877	Thomas Edison	Phonograph	The first recording machine. This one allowed a few minutes of sound effects to be engraved on a cylinder covered with a tin foil.
1886	Chester Bell and Charles Sumner Tainter	Graphophone	Improved invention of the phonograph.
1887	Emile Berliner	Gramophone	The cylinder was replaced with wax–coated zinc discs. This process also made it possible to create molds for industrial production of the cylinders. He also manufactured the first flat disc press and the apparatus for reading this type of media.
1889	Emile Berliner		Marketing of the first record players (gramophone) reading flat discs with a diameter of 12 cm.
1898	Emile Berliner		Foundation of the "Deutsche Grammophon Gesellschaft".
1898	Valdemar Poulsen	Telegraphone	First magnetic recording machine consisting of a piano wire unwinding in front of the poles of an electromagnet, whose current varied according to the vibrations of a sound-sensitive membrane.
1910			Standardization of the diameter of the recordable media: 30 cm for the large discs and 25 cm for the small ones. A few years later, normalization of the rotation speed of the discs at 78 rpm.

1	Inventor, author	Device	Description
1934	Marconi and Stille	Tape recorder	Recording machine on plastic tape coated with magnetic particles.
1948	Peter Goldmark		Launch of the first microgroove discs with a rotation speed of 33 rpm.
1958	"Audio Fidelity" Company	Release of the first stereo microgroove discs	
1980	Philips and Sony		Marketing of the CD (or compact disc), a small 12 cm disc covered with a reflective film.
1998			Implementation of MP3 compression.

Table 2.1. *Milestones in speech recording processes*[3]

2.3. Recording chain

The analog recording of speech sounds, used after the decline of mechanical recording but practically abandoned today, uses a magnetic coding of sound vibrations: a magnetic head consists of an electromagnet that modifies the polarization of the magnetized microcrystals inserted on a magnetic tape. The sound is reproduced by a similar magnetic head (the same head is used for recording and playback in many devices). The frequency response curve of this system, which characterizes the fidelity of the recording, depends on the speed of the tape, and also on the fineness of the gap between the two ends of the electromagnet in contact with the magnetic tape (a finer gap makes it possible to magnetize finer particles, and thus to reach higher recording frequencies). Similarly, a high speed allows the magnetized particles to remain between the gaps of the magnetic head for a shorter period of time, also increasing the frequency response.

Despite the enormous progress made both in terms of the magnetic oxides stuck to the tapes, and in terms of frequency compensation by specialized amplifiers and filters, magnetic recording is still tainted by the inherent limitations of the system. Today, some professionals occasionally use this type of recorder (for example, Kudelski's "Nagra"), with a running

3 http://www.gouvenelstudio.com/homecinema/disque.htm.

speed of 38 cm/s (the standard is actually 15 inches per second, in other words, 38.1 cm/s). Amateur systems have long used sub-multiple speeds of 15 inches, i.e. 19.05 cm/s, 9.52 cm/s and, in the cassette version, 4.75 cm/s.

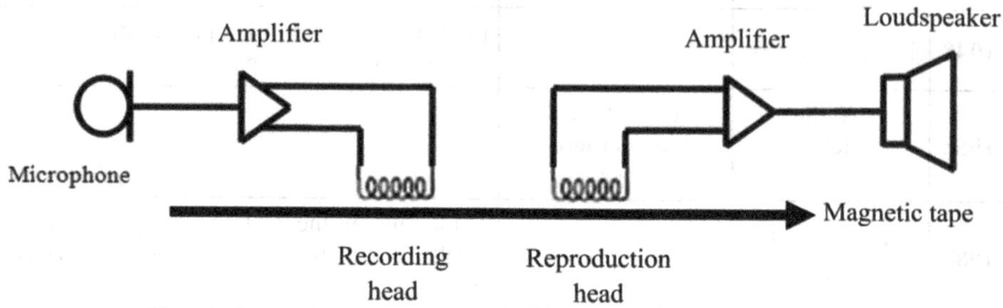

Figure 2.8. *Analog magnetic recording chain*

The width of the magnetic tape is related to the signal-to-noise ratio of the recorder. Early designs used 1-inch (2.54 cm) tapes, then came the half-inch and quarter-inch tapes for cassettes, on which two separate tracks were installed for stereophonic recording. Professional studio recorders used 2-inch and 4-inch tapes, allowing 8 or 16 simultaneous tracks. In addition to the physical limitations, there are distortions, due to wow and fluttering, caused by imperfections in the mechanical drive system of the tape: wow for slow variations in drive speed (the tape is pinched between a cylinder and a capstan), fluttering for instantaneous variations in speed.

2.3.1. *Distortion of magnetic tape recordings*

All magnetic recording systems are chains with various distortions, the most troublesome of which are:

– *Distortion of the frequency response*: the spectrum of the starting frequencies is not reproduced correctly. Attenuation occurs most often for low and high frequencies, for example, below 300 Hz and above 8,000 Hz for magnetic tapes with a low running speed (in the case of cassette systems, running at 4.75 cm/s).

– *Distortion of the phase response*: to compensate for the poor frequency response of magnetic tapes (which are constantly being improved by the use of new magnetic oxide mixtures), filters and amplifiers are used to obtain a better overall frequency response. Unfortunately, these devices introduce a deterioration of the signal-to-noise ratio and also a significant phase distortion for the compensated frequency ranges.

– *Amplitude*: despite the introduction of various corrective systems based on the masking effect (Dolby type), which compress the signal dynamics during recording and restore it during playback, the signal-to-noise ratio of magnetic tape recordings is in the order of 48 dB for cassette systems (this ratio is better and reaches 70 dB for faster magnetic tape speeds, for example, 38 cm/s for professional magnetic recorders) and is used by wider magnetic tapes (half-inch instead of quarter-inch).

– *Harmonics*: the imperfect quality of the electronic amplifiers present in the chain can introduce harmonic distortions, in other words, components of the sound spectrum that were not present in the original sounds.

– *Scrolling*: the mechanisms that keep the magnetic tape moving may have imperfections in their regularity during both recording and reproduction, resulting in random variations in the reproduced frequencies (the technical terms for this defect are wow and flutter, which represent, respectively, a slowing down and acceleration in the reproduction of sound). The magnetic tape stretching due to rapid rewinding can also produce similar effects.

– *Magnetic tape*: the thinness of magnetic tapes can be such that a section of the tape in the crosstalk can magnetize the parts wound up directly above or below in the crosstalk. The same effect can occur between two tracks that are too close together in the same section, for example, in stereo recording.

All these limitations, which require costly mechanical and electronic corrections, promote the outright abandonment of analogical recording systems. With the advent of low-cost SSD computer memories with information retention, the use of hard disks or, possibly, digital magnetic tapes now makes it possible to record very long periods of speech, with an excellent signal-to-noise ratio and a very good frequency response, the only weak links that remain in the recording chain being the microphone and the loudspeaker (or headphones), in other words, the converters from an analogical variation (sound pressure variations) to a digital variation, and vice versa.

2.3.2. *Digital recording*

All these shortcomings have given rise to a great deal of research of all kinds, but the development of SSD-type computer memories, on DAT magnetic tape or on hard disk, has completely renewed the sound recording processes.

By minimizing the use of analog elements in the recording chain, it is much easier to control the various distortions that could be introduced. In fact, only two analog elements remain in the recording chain: (1) the microphone, which converts pressure variations into electrical variations, and (2) the analog-to-digital converter, which converts sequences of numbers into an electrical analog signal, which will eventually be converted by a loudspeaker or earphone into sound pressure variations.

The digital recording chain (Figure 2.9) consists of a microphone, feeding an analog preamplifier, followed by an analog filter. This anti-aliasing filter (see below) delivers a signal, digitized by an analog-to-digital converter, a number of times per second, called the sampling rate. The signal is thus converted into a sequence of numbers stored in a digital memory of any type, SSD, disk (DAT tape is practically abandoned today). The reproduction of the digitized sound is done by sequentially presenting the stored numbers to a digital-to-analog converter which reconstructs the signal and, after amplification, delivers it to an earphone or a loudspeaker. The quality of the system is maintained through the small number of mechanical elements, which are limited to the first and last stages of the chain.

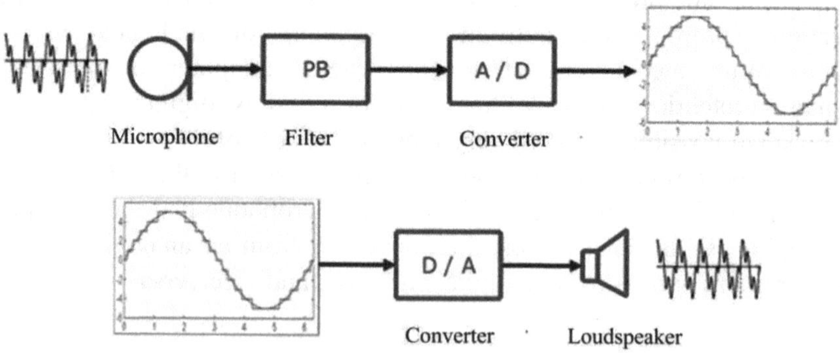

Figure 2.9. *Digital recording chain. For a color version of this figure, see www.iste.co.uk/martin/speech.zip*

2.4. Microphones and sound recording

There are many types of microphones: carbon, laser, dynamic, electrodynamic, piezoelectric, etc., some of which are of an old-fashioned design and are still used in professional recording studios. The principle is always the same: convert sound pressure variations into electrical variations. Dynamic microphones consist of an electromagnet whose coil, attached to a small diaphragm, vibrates with sound and produces a low electrical voltage, which must then be amplified. Condenser microphones (including electret microphones) use the capacitance variation of a material when exposed to sound vibration. Piezoelectric microphones use ceramic crystals that produce an electrical voltage when subjected to sound pressure.

All these sound pressure transducers are characterized by a response curve in amplitude and phase, and also by a polar sensitivity curve that describes their conversion efficiency in all directions around the microphone. The response curves are graphical representations of the detected amplitude and phase values as a function of the frequency of pure tones. By carefully selecting the type of response curve, be it omnidirectional (equal in all directions), bidirectional (better sensitivity forward and backward), or unidirectional (more effective when the source is located forward), it is possible to improve the sound recording quality, which is essentially the signal-to-noise ratio of the recording, with the noise corresponding to any sound that is not recorded speech (Figure 2.10).

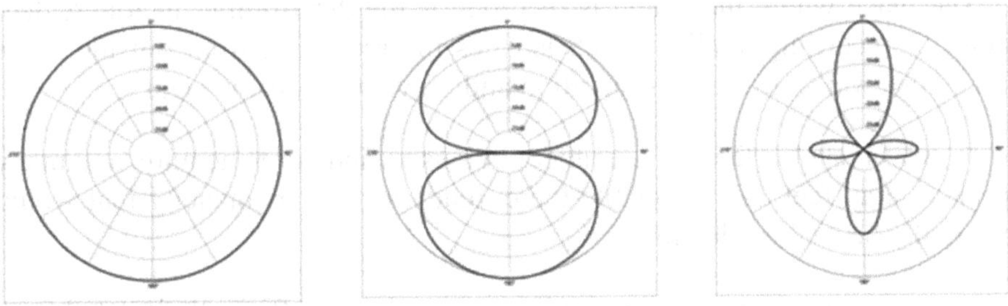

Figure 2.10. *Polar response curves of omnidirectional, bidirectional and unidirectional microphones*

There are also shotgun microphones, which offer very high directionality and allow high signal-to-noise ratio recordings at relatively long distances (5 to 10 meters) at the cost of large physical dimensions. This last feature requires an operator who constantly directs the shotgun microphone towards the sound source, for example, a speaker, which can be problematic in practice if the speaker moves even a few centimeters. Due to their cost, these microphones are normally reserved for professional film and television applications.

Today, most recordings in the media, as well as in phonetic research, use so-called lavalier microphones. These microphones are inexpensive and effective if the recorded speaker is cooperative (the microphones are often physically attached to the speaker), otherwise, unidirectional or shotgun microphones are used. Electret microphones require the use of a small bias battery, which, in practice, some tend to forget to disconnect, and which almost always proves to be depleted at the critical moment.

2.5. Recording locations

In order to achieve good quality, both in terms of frequency spectrum and signal-to-noise ratio (anything that is not speech sounds is necessarily noise), the recording must meet recommendations, which are common sense, but, unfortunately, often poorly implemented. Recording is the first element in the analysis chain, and one whose weaknesses cannot always be corrected afterward. The quality of the sound recording is therefore an essential element in the chain.

The place where the sound is recorded is the determining factor. A deaf room, or recording studio, which isolates the recording from outside noise and whose walls absorb reverberation and prevent echoes, is ideal.

However, such a facility is neither obvious to find nor to acquire, and many speakers may feel uncomfortable there, rendering the spontaneity that would eventually be desired fleeting.

In the absence of a recording studio, a room that is sufficiently isolated from outside noise sources may be suitable, provided that it has low reverberation (windows, tiles, etc.), and that no noise from dishes, cutlery, refrigerators, chairs being moved, crumpled paper, etc. is added to the recorded speech. Recording outdoors in an open space (and therefore not in the forest, which is conducive to echo generation) does not usually offer reverberation and echo problems, however, wind noise in the microphone could be a hindrance.

However, you can protect yourself from this (a little) by using a windscreen on the microphone. It will also be necessary to prevent any traffic noise or other noise, which is not always obvious.

The positioning of the microphone is also important: avoid the effects of room symmetry, which can produce unwanted echoes, place the microphone close to the speaker's lips (30 cm is an optimal distance) and provide mechanical isolation between the microphone stand and the table or floor that carries it (a handkerchief, a tissue, etc.), so that the microphone does not pick up the noise of the recorder motor or the computer cooling fan. You should also make sure that if the microphone is placed on a table, it is stable, and to allow enough distance between the legs of the table and those of a nervous speaker.

There are professional or semi-professional systems, such as boom microphones (used in film shooting) or lavalier microphones that are linked to the recording system by a wireless link (in this case, attention must be paid to the response curve of this link). The latter are often used in television: independence from a cable link allows greater mobility for the speaker, who must, obviously, be cooperative.

2.6. Monitoring

Monitoring the recording is essential. In particular, make sure that the input level is set correctly, neither too low (bad signal-to-noise ratio) nor too high (saturation). Almost all recording systems are equipped with a volume meter, which is normally scaled in dB, which allows you to view the extreme levels during recording.

In any case, it is absolutely necessary to refrain from using the automatic volume control (AVC setting) which, although practical for office applications, introduces considerable distortions in the intensity curve of the recording: the volume of sounds that are too weak is automatically increased, but with a certain delay, which also reinforces the background noise, which could then be recorded at a level that is comparable to that of the most intense vowels.

Ideally, real-time spectrograph monitoring should be available, allowing a user familiar with spectrogram reading to instantly identify potential problems, such as saturation, low level, echo, inadequate system response curve, and noise that would otherwise go unnoticed by the ear. The necessary corrections can then be made quickly and efficiently, because after the recording, it will be too late! Spectrographic monitoring requires the display of a spectrogram (normally in narrowband, so as to visualize the harmonics of the noise sources) on a computer screen, which is sometimes portable. Today, a few rare software programs allow this type of analysis in real time on PC or Mac computers (for example, WinPitch).

Figure 2.11. *A recording session with Chief Parakatêjê Krohôkrenhum (State of Pará, Brazil) in the field, with real-time monitoring of the spectrogram and the fundamental frequency curve (photo: L. Araújo). For a color version of this figure, see www.iste.co.uk/martin/speech.zip*

2.7. Binary format and Nyquist–Shannon frequency

The electrical signal delivered by the microphone must be converted, after adequate amplification, into a table of numbers. This is the digitization step. Two parameters characterize this so-called analog-to-digital conversion (ADC): the conversion format of the signal amplitude and the conversion frequency.

2.7.1. *Amplitude conversion*

Contemporary computers operate with binary digits: any decimal number, any physical value converted into a number is stored in memory and processed as binary numbers, using only the digits 0 and 1. Furthermore, computer memories are organized by grouping them into 8 binary digits, called bytes, making it possible to encode 255 states or 255 decimal numbers (in addition to the zero).

An analog signal such as a microphone signal has positive and negative values. Its conversion using a single byte therefore makes it possible to encode 127 levels or positive values (from 1 to 127) and 128 negative values (from −1 to −128). Intermediate values between two successive levels will be rounded up or down to the next higher or lower value, which introduces a maximum conversion error (also called quantization error) of 1/127, which in dB is equivalent to $20 \times \log (1/127) = 20 \times 2.10 = -42$ dB. In other words, conversion using a single byte introduces a quantization noise of −42 dB, which is not necessarily desirable.

Also, most analog-to-digital converters offer (at least) 10 or 12 binary digits (bits) conversion, which corresponds to conversion noises of $20 \times \log (1/511) = -54$ dB and $20 \times \log (1/1023) = -60$ dB, respectively. Since the price of memory has become relatively low, in practice, it is no longer even worth encoding each 12-bit value in 1.5 bytes (two 12-bit values in 3 bytes), and the 2-byte or 16-bit format is used for the analog-to-digital conversion of speech sound, even if the analog-to-digital conversion is actually done in the 12-bit format.

2.7.2. *Sampling frequency*

How many times per second should the analog variations be converted? If you set too high a value, you will consume memory and force the processor to pointlessly handle a lot of data, which can slow it down unnecessarily.

If too low a value is set, aliasing will occur. This can be seen in Figure 2.12, in which the sinusoid to be sampled has about 10.25 periods, but there are only 9 successive samples (represented by a square), resulting in an erroneous representation, illustrated by the blue curve joining the samples selected in the sampling process.

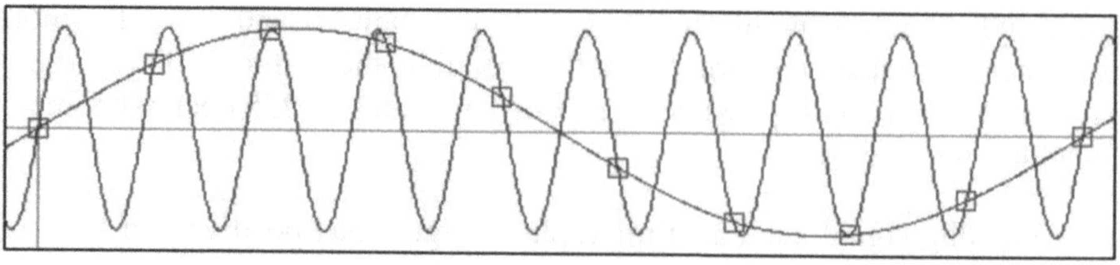

Figure 2.12. *Aliasing. Insufficient sampling frequency misrepresents the signal. For a color version of this figure, see www.iste.co.uk/martin/speech.zip*

The Nyquist–Shannon theorem (Harry Nyquist, 1889–1976 and Claude Shannon, 1916–2001) provides the solution: for there to be no aliasing, it is necessary and sufficient for the sampling frequency to be greater than or equal to twice the highest frequency (in the sense of the Fourier harmonic analysis) of the sampled signal.

This value is easily explained by the fact that at least two points are needed to define the frequency of a sinusoid, and that in order to sample a sinusoid of frequency f, at least double frequency, 2f, sampling is required.

The practical problem raised by Nyquist–Shannon's theorem is that one does not necessarily know in advance the highest frequency contained in the signal to be digitized, and that one carries out this conversion precisely in order to analyze the signal and to know its spectral composition (and thus its highest frequency).

To get out of this vicious circle, an analog low-pass filter is used, which only lets frequencies lower than half the selected sample rate pass through between the microphone and the converter. Higher-frequency signal components will therefore not be processed in the conversion and the Nyquist criterion will be met.

2.8. Choice of recording format

2.8.1. *Which sampling frequency should be chosen?*

The upper frequency of the speech signal is produced by the fricative consonant [s] and is around 8,000 Hz. Applying Nyquist–Shannon's theorem, one is led to choose a double sampling frequency, in other words, 2 × 8,000 Hz = 16,000 Hz, without worrying about the recording of occlusive consonants such as [p], [t] or [k], which are, in any case, mistreated in the recording chain, if only through the microphone, which barely satisfactorily translates the sudden pressure variations due to the release of the occlusions.

Another possible value for the sampling rate is 22,050 Hz which, like 16,000 Hz, is a value commonly available in standard systems. The choice of these frequencies automatically implements a suitable anti-aliasing filter, eliminating frequencies that are higher than half the sampling rate.

In any case, it is pointless (if one has the choice) to select values of 44,100 Hz or 48,000 Hz (used for the digitization of music) and even more so for stereo recording when there is only one microphone, therefore only one channel, in the recording chain.

2.8.2. *Which coding format should be chosen?*

In the 1990s, when computer memories had limited capacity, some digital recordings used an 8-bit format (only 1 byte per sample). As the price of memory has become relatively low, it is no longer even economically practical to encode each 12-bit value in two bytes (two 12-bit values in 3 bytes), and the 2-byte or 16-bit format is used for the analog-to-digital conversion of speech sounds.

2.8.3. *Recording capacity*

Using 2 bytes per digital sample and a sampling rate of 22,050 Hz, $2 \times 22{,}050$ bytes per second are consumed, in other words, $2 \times 22{,}050 \times 60 = 2{,}646{,}000$ bytes per minute, or $2 \times 22{,}050 \times 60 = 156{,}760{,}000$ bytes per hour, or just over 151 MB (1 Megabyte = $1{,}024 \times 1{,}024$ bytes), with most computer sound recording devices allowing real-time storage on a hard disk or SSD. A hard disk with 1,000 Gigabytes available can therefore record more than $151 \times 1{,}024 = 151{,}000$ hours of speech, in other words, more than 6,291 days, or more than 17 years!

2.9. MP3, WMA and other encodings

There are many methods for compressing digital files, and for generally allowing you to find the exact file that was originally compressed after decoding. On the other hand, for digitized speech signals, the transmission (via the Internet or cellular phones) and storage of large speech or music files has led to the development of compression algorithms that do not necessarily restore the original signal identically after decompression. The MP3 encoding algorithm belongs to this category, and uses the human perceptual properties of sound to minimize the size of the encoded sound files.

MP3 compression essentially uses two processes: 1) compression based on the ear-masking effect (which produces a loss of information), and 2) compression by the Huffman algorithm (which does not produce a loss of information).

Other compression processes exist, such as WMA, RealAudio or ATRAC. All of these systems use the properties of the frequency- (simulated sounds) or time- (sequential sounds) masking effect and produce an unrecoverable distortion of the speech signal, regardless of the parameters used (these methods have parameters that allow more or less efficient compression, at the cost of increased distortion for maximum compression).

While the compression standards are the result of (lengthy) discussions between members of specialist research consortia (MP2 and MP4 image compression, MP1 sound, etc.), the MP3 standard was patented by the Frauhofer laboratories. In reality, it is the MPEG–1 Layer 3 standard (the "layers" are classified by level of complexity), which a large number of researchers at the Frauhofer Institute have been working on defining, (MPEG is the name of a working group established under the joint leadership of the International Organization for Standardization and the International Electrotechnical Commission (ISO/IEC), which aims to create standards for digital video and audio compression).

There are algorithms that are optimized for audio signals and that perform lossless compression at much higher rates than widely-used programs, such as WinZip, which have low efficiency for this type of file. Thus, with the WavPack program, unlike MP3 encodings, the compressed audio signal is identical after decompression[4]. Other compression processes of this type exist, such as ATRAC, Advanced Lossless, Dolby TrueHD, DTS–HD Master Audio, Apple Lossless, Shorten, Monkey's Audio, FLAC, etc. The compression ratio is in the order of 50% to 60%, lower than those obtained with MP3 or WMA, for example, but the ratio obtained is without any loss of information. This type of compression is therefore preferable to other algorithms, as the information lost in the latter case cannot be recovered.

The Fraunhofer Institute/Thomson Multimedia Institute patent has been strictly enforced worldwide and the price of the licenses was such that many companies preferred to develop their own systems, which were also patented, but generally implemented in low-cost or free programs. This is the case with WMA compression (Microsoft), Ogg Vorbis (Xiph Org), etc. Today, the MP3 patent has fallen into the public domain and, furthermore, other improved standards compared to MP3 have emerged (for example: MP2–AAC, MP4–AAC, etc. (AAC, Advance Audio Coding)).

Before the expiry of the MP3 patents, some US developers found an original way to transmit MP3 coding elements, while escaping the wrath of the lawyers appointed by the Fraunhofer Institute: the lines of code were printed on T-shirts, a medium that was not mentioned in the list of transmission media covered by the patents. Amateur developers could then acquire this valuable information for a few dollars without having to pay the huge sums claimed by the Institute for the use of the MP3 coding process.

Detailed information and the history of the MP3 standard can be found on the Internet[5].

Box 2.1. *The MP3 code on T-shirts*

4 www.wavpack.com.

5 http://www.mp3–tech.org/.

3

Harmonic Analysis

3.1. Harmonic spectral analysis

As early as 1853, Scott de Martinville had already examined the details of vowel vibrations inscribed by his phonautograph on smoke-blackened paper, using a magnifying glass. He had noted that the classification of speech sounds, and vowels in particular, did not seem to be possible on the basis of their representation in waveform because of the great variations in the observed patterns. Figure 3.1 illustrates this problem for four realizations of [a] in the same sentence by the same speaker.

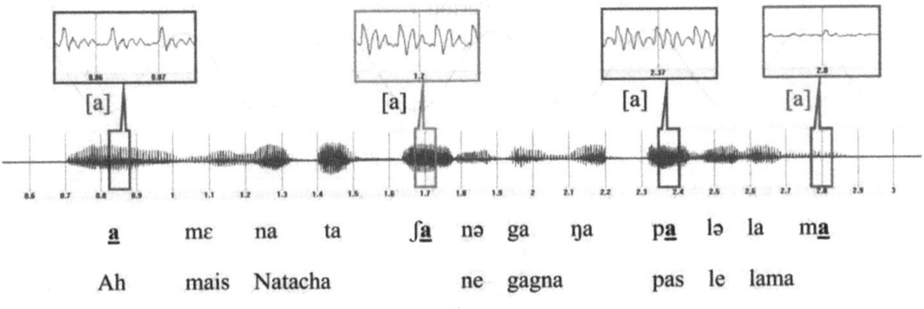

Figure 3.1. *Four realizations of [a] in stressed syllables of the same sentence in French, showing the diversity of waveforms for the same vowel. The sentence is: "Ah, mais Natasha ne gagna pas le lama" [amɛnataʃanəgaɲapaləlama] (voice of G.B.)*

By adding pure harmonic sounds – in other words, integer multiples of a basic frequency called the fundamental frequency – and by cleverly shifting

For a color version of all the figures in this chapter, see: www.iste.co.uk/martin/speech.zip.

the harmonics with respect to each other – that is by changing their respective phases – we can obtain a so-called complex waveform that adequately resembles some of the observed vowel patterns. This is because phase shifts cause significant changes in the complex waveform at each period (Figure 3.2). The trick is to find a method for calculating the amplitudes and phases of each harmonic component: this is Fourier analysis.

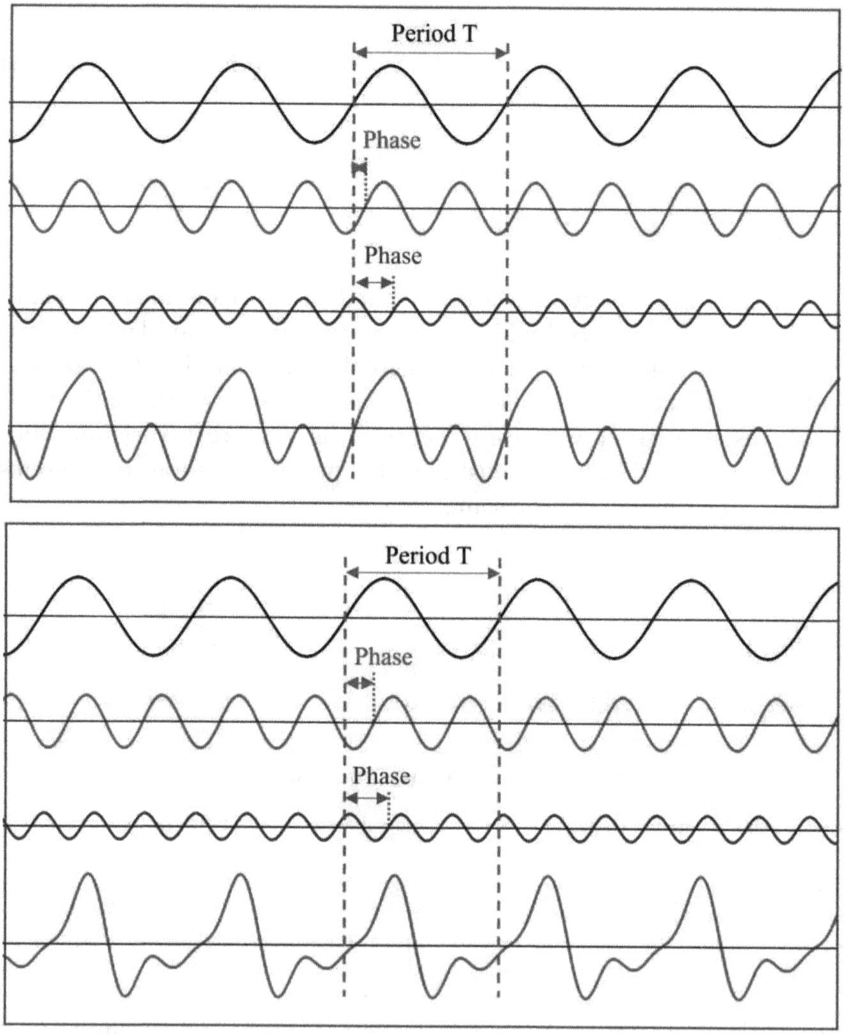

Figure 3.2. *Effect of the phase change of three harmonics on the waveform, resulting from the addition of components of the same frequency but different phases at the top and bottom of the figure*

Fourier harmonic analysis, known since 1822, provides a method of analysis used to describe speech sounds in a more efficient way. This is done by performing the opposite operation of the additions illustrated in Figure 3.2; in other words, by decomposing the waveform into a series of pure harmonic sounds, the addition of which (with their respective phases) restored the original waveform. The idea of decomposing into trigonometric series seems to have already appeared in the 15th Century in India, and would be taken up again in the 17th and 18th Centuries in England and France for the analysis of vibrating strings. In fact, Fourier analysis applies to periodic functions that are infinite in time, and it will therefore be necessary to adapt this constraint to the (hard) reality of the signal, which, far from being infinite, changes continuously with the words of the speaker.

To solve this problem, the idea is to sample segments of the sound signal at regular intervals, and to analyze them as if each of these segments were repeated infinitely so as to constitute a periodic phenomenon; the period of which is equal to the duration of the sampled segment. We can thus benefit from the primary interest of Fourier analysis, which is to determine the amplitude of the harmonic components and their phase separately. This is in order to obtain what will appear as an invariant characteristic of the sound with the amplitude alone, whereas the phase is not relevant, and only serves to differentiate the two channels of a stereophonic sound, as perceived by our two ears.

The principle of harmonic analysis is based on the calculation of the existing correlation between the analyzed signal and two sinusoidal functions, offset by 90 degrees (π/2); in other words, a correlation with a sine and a cosine. The modulus (the square root of the sum of the squares) of the two results will give the expected response, independently of the phase, which is equal to the arc of the tangent of the ratio of the two components. Mathematically, the two components A and B, of the decomposition of the sampled signal of duration T, are obtained by the following equations:

$$A = \frac{2}{T} \sum_{n=-T/2}^{T/2} \mathrm{f}(n/T) \cos(2\pi n/T)$$

$$B = \frac{2}{T} \sum_{n=-T/2}^{T/2} \mathrm{f}(n/T) \sin(2\pi n/T)$$

in other words, the sum of the values taken from the beginning to the end of the speech segment sampled for the analysis, multiplied by the corresponding cosine and sine values at this frequency F, with F = 1/T. The amplitude of the sine resulting from this calculation is equal to $\sqrt[2]{(A^2 + B^2)}$ for this frequency, and its phase is equal to arc tg(B/A) (the value of the angle whose tangent is equal to B/A).

In reality, beneath this apparently complicated formula lies a very simple mathematical method of analysis: correlation. Correlation consists of the multiplication of the signal or part of the signal by a function of known characteristics. If there is a strong similarity between the analyzed function and the correlation function, the sum of the products term to term – the integral in the case of continuous functions, the sum of the products of the corresponding samples in the case of digitized functions – will be large, and this sum will be low in the case of weak correlation.

Fourier series analysis proceeds in this way, but a problem arises because of the changing phases of the harmonic components. To solve this, two separate correlations are actually performed, with sinusoidal functions shifted by 90 degrees (π/2). By recomposing the two results of these correlations, we obtain the separate modulus and phase (Figure 3.3).

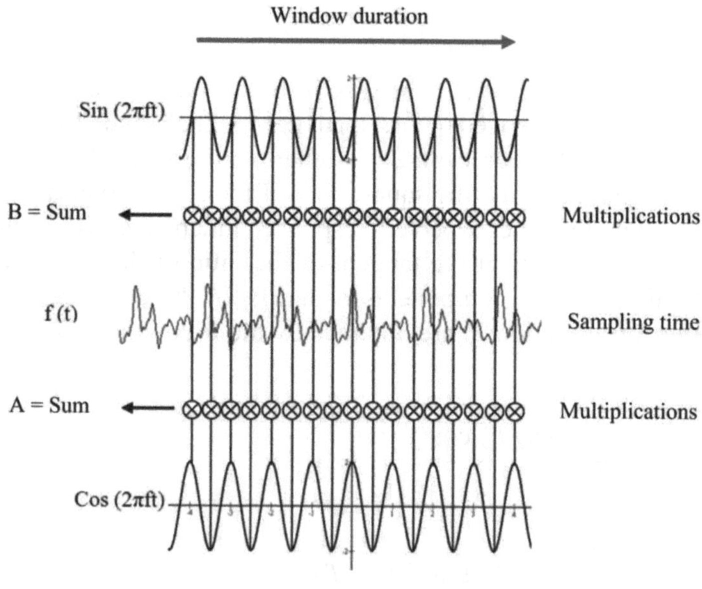

Figure 3.3. *Diagram of the principle of harmonic analysis in Fourier series*

Already, at the beginning of the 20th Century, experimental plots were sampled graphically to obtain Fourier coefficients, which were then plotted with the amplitude on the ordinate and the frequency on the abscissa, so as to facilitate the interpretation of the results. This graphical representation is called the amplitude spectrum (see Figure 3.4). A graph plotting phase versus frequency is therefore called a phase spectrum.

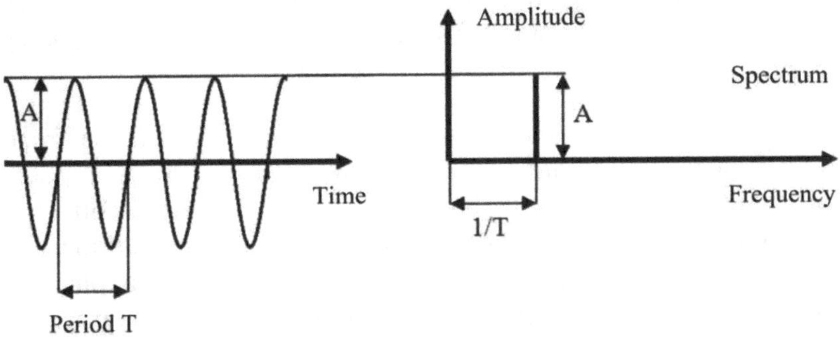

Figure 3.4. *Correspondence between the temporal (left) and frequency (right) representation of a pure sound of period T and amplitude A*

Fourier harmonic analysis therefore consists of multiplying the signal samples term by term using values sampled at the same sine and cosine instants, and adding the results for all samples within the time window (Figure 3.3). This analysis requires a long series of multiplications and additions, which nowadays is performed rapidly by computer. At that time, each value sampled and measured by hand had to be multiplied by sine and cosine values obtained from a table. Then, for each frequency, the results of these multiplications had to be added together and the modulus of the sine and cosine series were calculated. This was tedious work which required weeks of calculation, and which was sometimes subcontracted in monasteries at the beginning of the 20th Century.

There is a price to pay (there is always a price) for operating on signal segments of limited duration T (T for "Time") and that are not infinite, as imposed by the Fourier transform, and imagining that the segment reproduces itself from – infinity to + infinity in a periodic manner, with a period equal to the duration T of the segment (Figure 3.5). Since the analysis amounts to decomposing a signal that has become periodic with a period T, the harmonics resulting from this decomposition will have frequencies that are multiples of the basic frequency; in other words, 1/T, the reciprocal of the period. The frequency resolution, and thus the spacing of the components on the frequency axis, is therefore inversely proportional to the duration of the segments taken from the signal.

To obtain a more detailed spectrum, a longer duration of the sampled segment is therefore required. Consequently, the spectrum obtained will describe the harmonic structure corresponding to all the temporal events of the segment, and therefore any possible change in laryngeal frequency that may occur.

A longer speech segment duration gives us a more detailed frequency spectrum, which will only inform us about the "average" frequency structure of the sampled time segment, but will provide nothing at all about its possible evolution within the segment. Everything happens as if the analyzed sound was frozen for the time of the sample, just like a photographic snapshot is a frozen and sometimes blurred representation of reality.

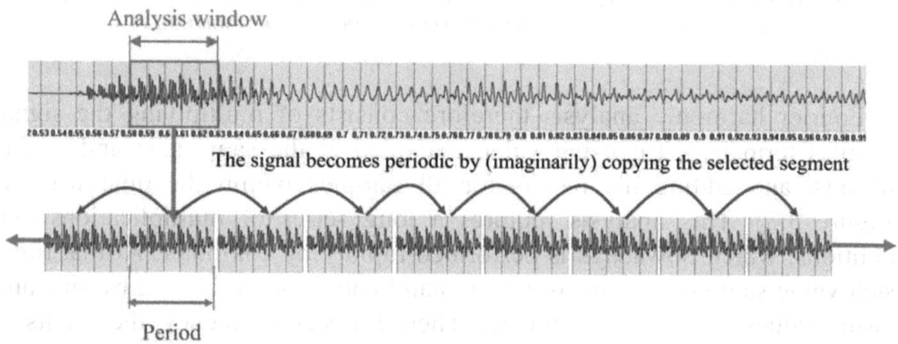

Figure 3.5. *Transformation of a sampled segment into a periodic signal*

The fundamental frequency (which has nothing to do *a priori* with the laryngeal vibration frequency) of this periodic signal is equal to the opposite of its period, thus the duration of the sampled segment. The longer the duration of the segment, the smaller the fundamental frequency will be (the frequency being the opposite of the period F = 1/T), and thus more details of the spectrum will be obtained. Conversely, a sample of shorter duration will correspond to a larger fundamental Fourier frequency, and thus a less detailed frequency spectrum.

This is known as a discrete Fourier spectrum because it consists of amplitude values positioned at frequencies that are multiples of the fundamental frequency. The discrete spectrum actually corresponds to the sampling of the continuous Fourier spectrum at intervals equal to 1/T (Figure 3.6).

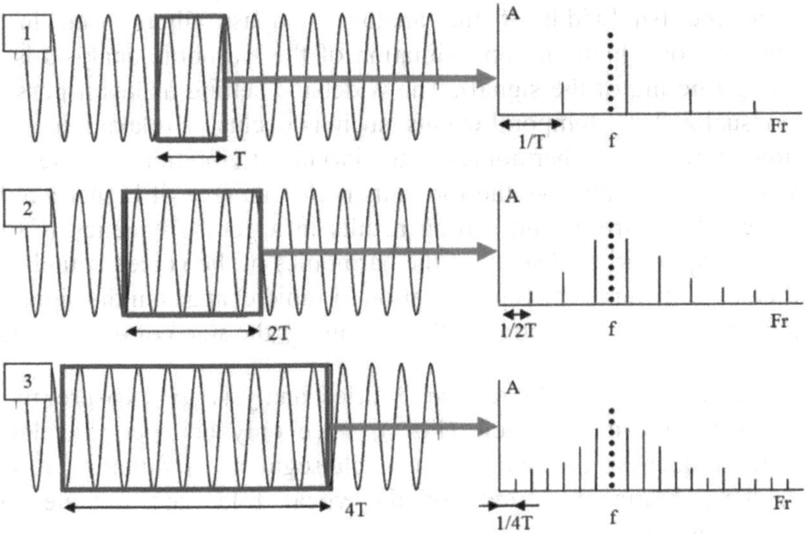

Figure 3.6. *Increase in frequency resolution with duration of the time window*

Figure 3.6 illustrates the interdependence of the frequency resolution with the duration of the time window. When this duration is T, the frequency resolution, i.e. the spacing between two consecutive values of the frequency in the spectrum (case 1), is equal to 1/T. When the sampling time is 2 T, a frequency spacing in the spectrum of 1/2 T is obtained, in other words, twice the frequency resolution (case 2). Lastly, when the duration is 4 T, the frequency resolution is doubled again, reaching 1/4 T (case 3), which reduces the error made on the frequency estimate of the analyzed sound (here, a pure sound of frequency equal to 7/8 T is represented by a dotted line). However, the price of a better frequency resolution is the loss of spectral change details over time, since each segment freezes the possible changes in the spectrum that may occur there. Even when using segments that are very close together on the time axis, each spectrum will only correspond to a kind of average of the harmonic variations of the signal within each segment.

This is reflected in the uncertainty principle, which often appears in theoretical physics, that one cannot win on both duration and frequency, which are in fact the opposite of one another. High precision on the time axis is paid for by low frequency resolution, and high time precision is paid for by low frequency resolution. This is the reason for the so-called “wideband” and “narrowband” settings of the first analog spectrographs (the term “band” refers to the bandwidth of the analog bandpass filters used in these instruments, for which an approximation of the harmonic analysis is made by analog filtering of the signal). The wideband setting makes it possible to better visualize brief temporal events, such as occlusive relaxation, and also to deliberately blur the harmonics of the laryngeal vibrations for the vowels, in order to visually observe the formants better, an area of higher amplitude harmonics. The narrowband setting results in a good frequency resolution and thus an appropriate display of the harmonics of the voiced sounds, but at the price of blurring in the representation of rapid changes in the signal, such as occlusive triggers or the onset (the beginning, the start) of the voicing.

It is also not possible to use Fourier analysis to measure the fine variations from cycle to cycle (jitter), since only one period value, and therefore frequency, is obtained for each segment of duration T, during which the vibrational periods of the vocal folds are not necessarily completely invariant.

The “right” duration of temporal sampling will depend on the signal analyzed and, in particular, on the duration of the laryngeal cycle during temporal sampling. For example, for adult male speakers, whose laryngeal frequency typically varies within a range of 70 Hz to 200 Hz, (thus a cycle time ranging from 14.8 ms to 5 ms) a duration of at least 15 ms is adopted so that at least one cycle is contained within the sampled segment. For an adult female voice, ranging for example from 150 Hz to 300 Hz, (thus from 6.6 ms to 3.3 ms) a value of 7 ms is chosen, for example.

One might think that for Fourier analysis of voiced sounds it would be ideal to adopt a sampling time equal to the duration of a laryngeal cycle. The harmonics of the Fourier series would then correspond exactly to those produced by the vibration of the vocal folds. The difficulty lies in measuring this duration, which should be done before the acoustic analysis. However, this could be achieved, at the expense of additional spectrum calculations, by successive approximations merging towards a configuration where the duration of the analyzed segment corresponds to one laryngeal period (or a sub-multiple).

Thus, a commonly used 30 ms sample corresponds to a Fourier fundamental frequency (never to be confused with the fundamental frequency estimate of the laryngeal frequency) of 1/30 ms = 33.3 Hz, barely sufficient to estimate the laryngeal frequency of a speech segment. However, during these 30 ms, the laryngeal frequency of 100 Hz, for example, (i.e. a laryngeal cycle duration of 10 ms), has the time to perform three cycles, and therefore to vary from cycle to cycle (known as the jitter in the physiological measurement of phonation), by 2%, for example, that is to say, varying from 98 Hz to 102 Hz. The Fourier series will hide this information from us and provide (by interpolation of the amplitude peaks of the harmonic components of the spectrum) a value of 100 Hz. Conversely, a duration of the sampled segment corresponding exactly to the duration of a laryngeal cycle in our example, i.e. 10 ms, will give a spacing of 1/10 ms = 100 Hz to the harmonics of the Fourier spectrum, with each of these harmonics coinciding with those of the sampled speech segment.

3.2. Fourier series and Fourier transform

The sum of sine and cosine functions resulting from the analysis of a speech segment, digitized according to the formulas:

$$A = \frac{2}{T}\sum_{n=-T/2}^{T/2} \mathrm{f}(n/T)\cos(2\,\pi t\, n/T)$$

and $B = \frac{2}{T}\sum_{n=-T/2}^{T/2} \mathrm{f}(n/T)\sin(2\,\pi t\, n/T)$

is a Fourier series. If the speech segments are not sampled and are represented by a continuous function of time, it is called a Fourier transform. The summations of the previous formulas are replaced by integrals:

$$A = \int_{-\infty}^{+\infty} f(t)\cos(2\pi t) \quad B = \int_{-\infty}^{+\infty} f(t)\sin(2\pi t)$$

with the module $\sqrt[2]{(A^2 + B^2)}$ and the phase arc $\mathrm{tg}(B/A)$.

3.3. Fast Fourier transform

When you have the curiosity (and the patience) to perform a Fourier harmonic analysis manually, you quickly realize that you are constantly performing numerous multiplications of two identical factors to the nearest sign (incidentally, the monks who subcontracted these analyses at the beginning of the 20th Century had also noticed this). By organizing the calculations in such a way as to use the results – several times over – of multiplications already carried out, a lot of time can be saved, especially if, as at the beginning of the 20th Century, there is no calculating machine. However, in order to obtain an optimal organization of the data, the number of values of the signal samples to be analyzed must be a power of 2, in other words, 2, 4, 8, 16, 32, 64, 128, 256, 512, 1,024, 2,048, 4,096, etc., in order to obtain the best possible organization of the data.

These observations were used by Cooley and Tuckey in 1965 to present a fast Fourier transform (FFT) algorithm, which takes advantage of the recurrent symmetry of the harmonic analysis calculations. Instead of the necessary $2N^2$ multiplications, only $N\log 2\ (N)$ multiplications are needed. Thus, a time sample of 1,024 points, corresponding to a duration of 64 ms with a sampling rate of 16,000 Hz, requires 2 × 1 024 × 1,024 = 2,097,152 multiplication operations for a discrete transform, whereas the fast transform requires only 1,024 × 10 = 10,240!

The number of frequency values obtained by FFT for the spectrum is optimal and equal to half the number of samples. The disadvantage is that the number of samples must be processed to the power of 2, but if the number of samples is between two powers of 2, zero values are added to obtain the desired total number. On the other hand, calculating the discrete Fourier transform (DFT) makes it possible to calculate the amplitude and phase of any frequency (less than the Nyquist frequency, of course, which is half the sampling frequency), and of any number of successive samples.

3.4. Sound snapshots

Phonation is the result of continuous articulatory gestures on the part of the speaker. So how do you make an acoustic analysis of these continuous movements? The principle is the same as the one used in cinema: if the speed of movement is not too high, a photographic snapshot is taken 24 times per second (in television, it is taken 25 or 30 times per second). For events that change more quickly, for example to film an athlete running 100 meters the number of snapshots per second is increased. Filming the vibration cycles of the vocal folds requires even more frames per second (2,000 frames per second is a commonly adopted value), since the duration of a vibration cycle can be as short as 10 ms at 100 Hz.

Not every snapshot is actually instantaneous. In photography, it takes a certain amount of exposure time to mark the light-sensitive film, or the light-sensitive diode array in digital photography. Acoustic analysis of speech and film recording present similar aspects: articulatory movements during speech production are gestures whose speed of establishment is not very different from that of other human gestures such as walking, for example. In order to be able to use acoustic analysis techniques based on periodic events, in other words, by stationary hypothesis, we will take sound "snapshots" by sampling a segment of the sound signal a certain number of times per second, for example 30 times – a figure comparable to the 24 frames per second in cinema – and then make a spectral analysis of it.

The hypothesis of periodicity and stationarity of Fourier analysis is obviously not at all valid in the construction of a periodic wave by reduplication of samples of the speech signal. It is an approximation intended to be used with mature mathematical methods that have benefited from much research and improvement in their practical implementation. Due to the relatively slow development of mathematical methods for describing essentially non-stationary events such as phonation, the tradition continues and the Fourier and Prony methods (see Chapter 5) remain the basis of modern acoustic speech analysis today.

3.5. Time windows

However, there is a little more to it than that! To see this, let us consider the acoustic analysis of a pure sound. As we know, pure sound is described mathematically by a sinusoid that is infinite in time. Very logically, the Fourier analysis in a sum of pure harmonic sounds should give a single spectral component, with a frequency equal to that of the analyzed pure sound, and a phase corresponding to the origin of the eventual defined time.

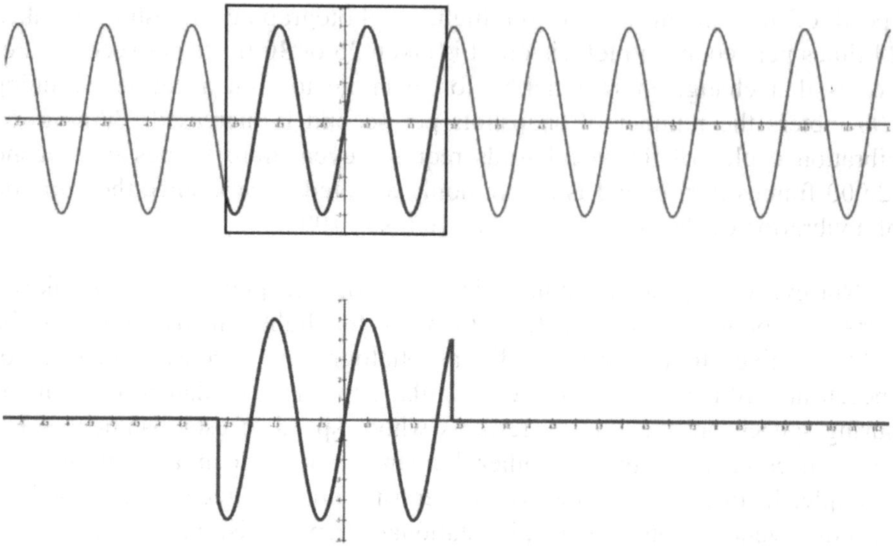

Figure 3.7. *Time sampling of pure sound through a rectangular window*

What happens when you isolate a segment within the pure sound? Unless you are very lucky, or know in advance the period of the pure sound being analyzed, the sampling time will not correspond to this period. By sampling and (mathematical) reproduction of the segment to infinity, we will have transformed another sound – no longer really described by a sinusoid but rather by a truncated sinusoid at the beginning and end (Figure 3.7).

Fourier analysis of this newly modified signal will result in a large number of parasitic harmonic components, foreign to the frequency of the original pure sound. Only when the duration of the time window corresponds exactly to the duration of one period of the pure sound, will the Fourier spectrum show one component alone.

So, how is this achieved? The example of sampling through a rectangular window illustrates the problem: it is intuitively understood that it is the limits of the window that cause these unwanted disturbances in the spectrum, through the introduction of artifacts into the sound being analyzed. We can then make them less important, by reducing their amplitude, so that the beginnings and ends of the sampled signal count less in the calculation of the Fourier spectrum because they have less amplitude. Optimal "softening" of both ends of the window is an art in itself, and has been the subject of many mathematical studies, where the Fourier transform of a pure sound is calculated to evaluate the effect of the time window on the resulting spectrum.

3.6. Common windows

To minimize the truncation effect caused by windowing, a large number of "softening" windows have been proposed. The most commonly used are:

– *the rectangular window*: the simplest, which gives the most selective top of the spectrum but also the most important rebounds in amplitude. It is the only window that considers all the information contained in the signal, since the amplitude of the signal is not modified anywhere inside the window;

– *the cosine window*: defined by the mathematical formula:

$$w(n) = \cos\left(\frac{\pi n}{N-1} - \frac{\pi}{2}\right) = \sin\left(\frac{\pi n}{N-1}\right)$$

– *the triangular window*: defined by:

$$w(n) = \frac{2}{N-1}\left(\frac{N-1}{2} - \left|n - \frac{N-1}{2}\right|\right)$$

– *the Blackman-Harris window*: whose equation is:

$$w(n) = a_0 - a_1 \cos\left(\frac{2\pi n}{N-1}\right) + a_2 \cos\left(\frac{4\pi n}{N-1}\right) - a_3 \cos\left(\frac{6\pi n}{N-1}\right)$$

– *the Hann(ing) window*: the most used but not necessarily the best for phonetic analysis, defined by:

$$w(n) = 0.5\left(1 - \cos\left(\frac{2\pi n}{N-1}\right)\right)$$

Figure 3.8 compares the spectra of pure sound at 1,500 Hz, obtained from several windows with an equal duration of 46 ms.

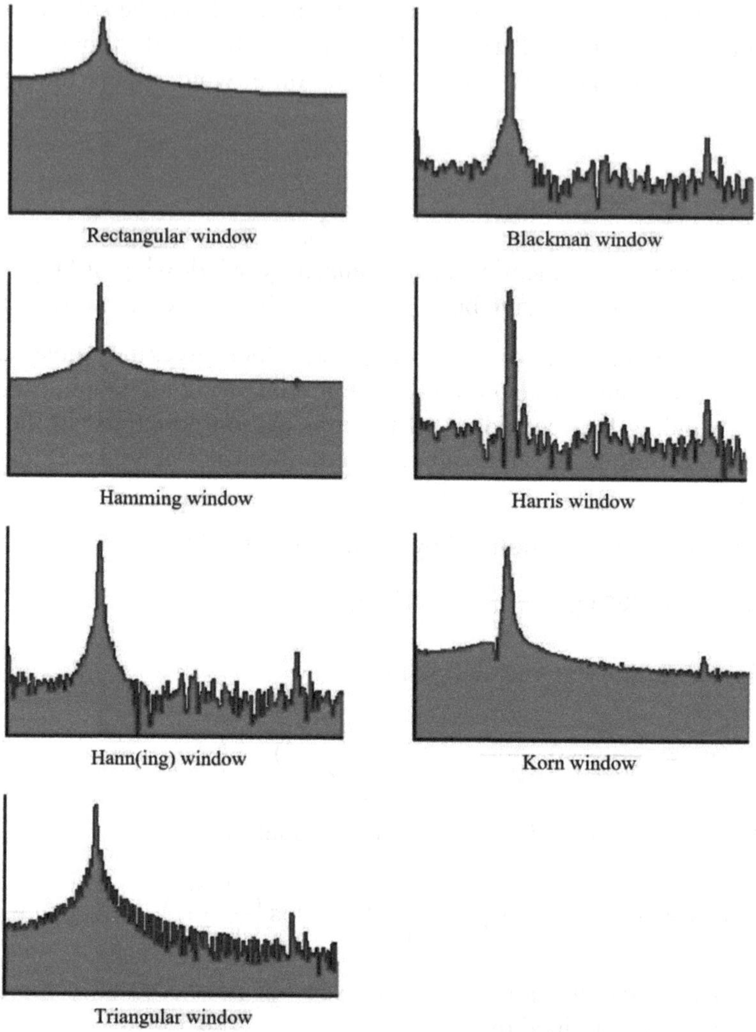

Figure 3.8. *Spectrum of pure sound at 1,500 Hz, 512 dots, 46 ms seen through different windows*

Each window has advantages and disadvantages. The Harris window has the best ratio of harmonic peak intensity and width, but the Hann(ing) window is still the most widely used (Figure 3.9).

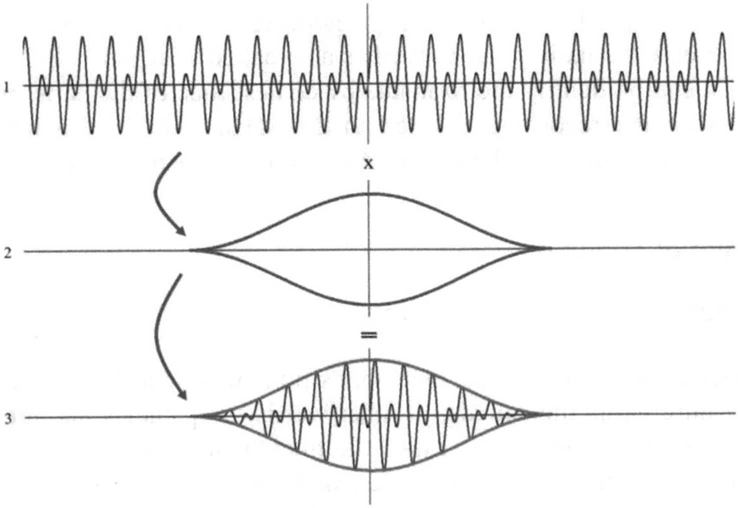

Figure 3.9. *Time sampling of the speech signal through a Hann(ing) window. The sampled signal (3) results from the multiplication of the signal (1) by the window (2)*

3.7. Filters

A filter is a device that attenuates or suppresses certain component frequencies of the signal. There are low-pass filters, which, as the name suggests, allow frequencies below a value known as the cut-off frequency to pass through, and attenuate and eliminate higher frequencies; high-pass filters, which eliminate low frequencies and allow frequencies above their cut-off frequency to pass through; and band-pass filters, which only allow frequencies between two limits to pass through.

The filters are either made by electronic components in the field of analog signal processing or by algorithms operating on the digitized signal values. In their actual implementation, they not only introduce a change in the amplitude spectrum of the filtered signal, but also in the phase spectrum, which is generally less desirable. Thus, in a low-pass filter, signal components of frequencies close to the cut-off frequency may be strongly out of phase, which can be problematic in the case of speech filtering, such as in the case of a speech attack: by low-pass filtering at 1,000 Hz, components in the 100–200 Hz range will come out of the filter after those in the 900–1,000 Hz range! Analog spectrographs such as Kay Elemetrics or Voice ID. avoided this problem by always using the same filters: a so-called narrowband filter (45 Hz), to obtain a good temporal resolution and to be able to observe the harmonics; as well as a so-called wideband filter (300 Hz) to better visually identify the formants by coalescence of the harmonics, and by analyzing the recording (limited to 2.4 s!) modified by a heterodyne system that was similar to the one used for radio receivers of the time.

3.8. Wavelet analysis

3.8.1. *Wavelets and Fourier analysis*

Fourier analysis allows the decomposition of a speech segment into a series of pure harmonic sounds, in other words, frequencies that are integer multiples of a basic frequency, called the fundamental frequency. The pitfall caused by the essentially non-periodic character of the speech signal for this type of analysis is overcome by the use of a time window that "freezes" the variations of the signal during the duration of the window, effectively transforming the non-periodic into the periodic.

The side effects of temporal windowing are known: (1) generation of frequency artifacts by the shape of the window, in particular by the need to have a null value at its extremities to limit its duration, and (2) opposite frequency and time resolution, leading to the choice of a compromise between the duration of the window and the interval between two harmonics, i.e., the value of the Fourier fundamental frequency.

However, in speech analysis, the spectral frequency range normally extends from 70 Hz to 8,000 Hz. In the lower frequency range, below 500 Hz or 800 Hz, it is often the laryngeal frequency that is of most interest, whereas for the higher frequencies of the spectrum, it is the formant frequencies, i.e., the frequencies of the areas where the harmonics have a greater amplitude (see Chapter 6). The ideal would therefore be to obtain spectra whose frequency resolution changes with frequency: good resolution for low frequencies, at least better than half the laryngeal frequency, and less good frequency resolution for higher frequencies, in the region of the formants, resulting in better temporal resolution.

Since the frequency resolution is inversely proportional to the duration of the analysis time window, such a configuration implies a variable duration of this window. This variation can be arbitrary, but to remain within the domain of Fourier harmonic analysis, the simplest way is to consider a duration directly related to the period (and thus the frequency) of the sine and cosine functions and, even more simply, to their number of periods, which is the basic principle of wavelet analysis. Instead of having a window of fixed duration (thus linked to the frequency resolution, which is also fixed) that determines the number of cycles of the periodic analysis functions, it is the number of cycles of these functions that determines the duration of the time window (Figure 3.10).

Wavelet analysis of a time signal is in fact a generalization of this principle. Instead of being limited to sine and cosine periodic functions, other periodic functions (oscillating functions) can be chosen, modulated by an appropriately shaped time window. The duration of the window will always be determined by the number of cycles of the periodic analysis function, the number of cycles having to be an integer (zero integral condition).

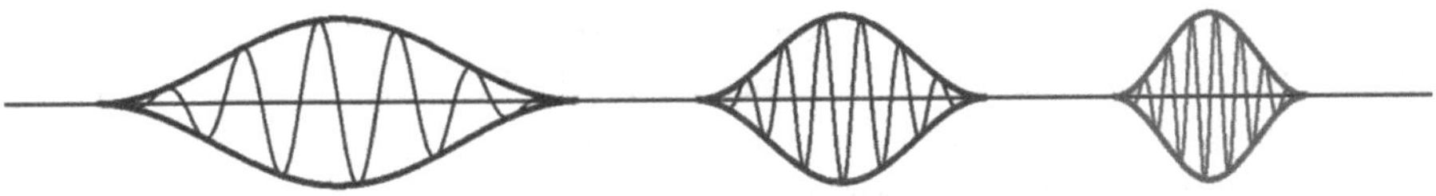

Figure 3.10. *Wavelet analysis window, 6 cycles at 3 different frequencies corresponding to different window durations*

3.8.2. *Choice of the number of cycles*

We have seen that the duration of the analysis window determines the frequency resolution. As in Fourier analysis, an amplitude-frequency spectrum is obtained by calculating the correlations between a speech segment that is extracted from the signal via a time window, and sine and cosine analysis functions oscillating at the desired frequencies. These analysis frequencies need not be integral multiples of the opposite of the window duration, but all intermediate values are in fact interpolations between two values which are multiples of the opposite of the window duration, interpolations which do not improve the frequency resolution. It is therefore more "economical", from a computational point of view, to choose analysis frequencies that are integer multiples of the fundamental frequency (in the Fourier sense), and to (possibly) proceed to a graphical interpolation on the spectrum obtained. The fast Fourier transform (FFT) algorithm proceeds in this way, the harmonics obtained essentially being integer multiples of the reciprocal of the duration of the analysis window.

Unlike Fourier, wavelet analysis does not proceed with constant frequency and time resolution. We therefore have to choose a desired value for a certain frequency.

Let us suppose we want to obtain a frequency resolution of 10 Hz for the 100 Hz frequency, in order that we can clearly distinguish the harmonics of the laryngeal frequency on the wavelet spectrum. The duration of the time window should therefore be 1/10 = 100 ms. Wavelet analysis imposes a zero average for the analysis function, which implies an integer value of the number of cycles for the analysis function. It will therefore take 10 cycles for the sine and cosine functions of the wavelet at 100 Hz for the duration of the wavelet, and therefore of the analysis window, to be 100 ms.

Once the (integer) number of cycles of the wavelet has been fixed for a frequency in the spectrum, the frequency and time resolutions relative to the other analysis frequencies will be derived. For our example, with a resolution of 10 Hz to 100 Hz, the 10 cycles of the wavelet at 200 Hz involve a window duration of 10 times 1/200 = 50 ms and a frequency resolution of 1/0.050 = 20 Hz; and at 1,000 Hz a window duration of 10 ms and a frequency resolution of 1/0.010 = 100 Hz, thus a value every 100 Hz, which is appropriate for the observation of formants on a spectrogram, all while obtaining a good resolution at 100 Hz for the visualization of the laryngeal frequency.

So, in one spectral analysis, we have appropriate settings that require several spectra by Fourier analysis. In reality, wavelet analysis is little used in the field of speech, but appears more frequently in the analysis of brain waves by electroencephalography (EEG), in a frequency range from 0.5 Hz to 50 Hz, and a color display whose code allows a better evaluation of the amplitudes of the components of the spectrum.

The following figures illustrate this property, including the variation of frequency resolution according to frequency, with wavelets of 5, 10 and 15 cycles. On a three-dimensional representation called a spectrogram (see Chapter 6), with time on the abscissa, frequency on the ordinate and the amplitude of the spectral components being color-coded, the range of good frequency resolution can be seen to increase from a window of 5 to 15 cycles.

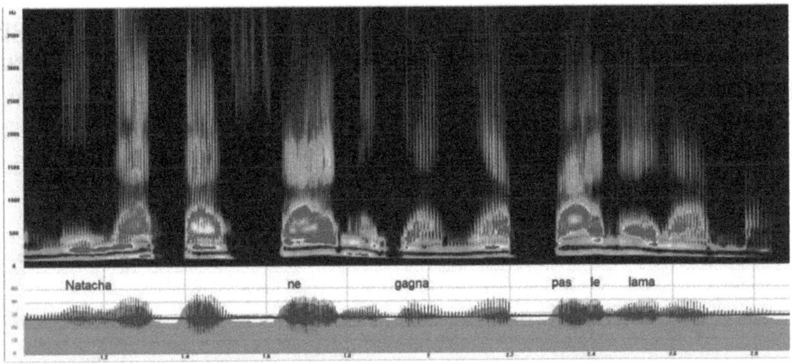

Figure 3.11. *Wavelet spectrogram, 5 cycles*

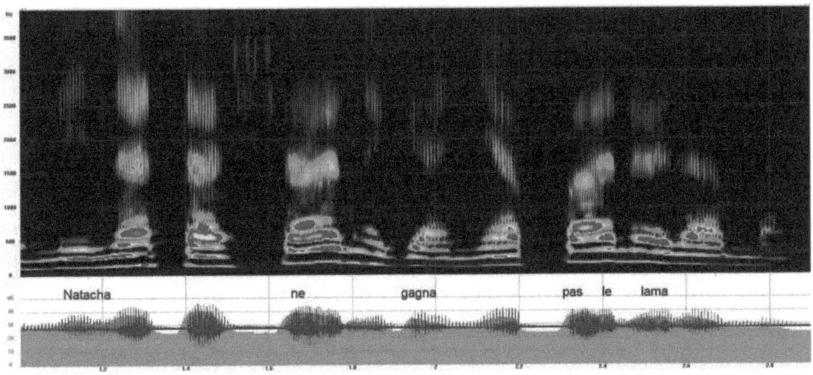

Figure 3.12. *Wavelet spectrogram, 10 cycles*

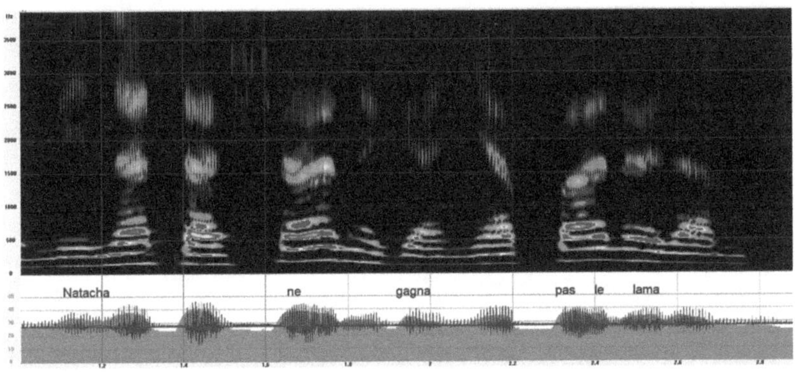

Figure 3.13. *Wavelet spectrogram, 15 cycles*

Fast wavelet transform (FWT) is a mathematical algorithm designed to transform a waveform or signal, in the time domain into a sequence of coefficients that are based on an orthogonal basis of finite small waves, or wavelets. This algorithm was introduced in 1989 by Stéphane Mallat.

4

The Production of Speech Sounds

4.1. Phonation modes

There are four ways of producing the sounds used in speech, and therefore four possible sources:

1) by the vibration of the vocal folds (the vocal cords), producing a large number of harmonics;

2) by creating turbulence in the expiratory airflow, by means of a constriction somewhere in the vocal tract between the glottis and the lips;

3) by creating a (small) explosion by closing the passage of exhalation air somewhere in the vocal tract, so as to build up excess pressure upstream and then abruptly releasing this closure;

4) by creating a (small) implosion by closing off the passage of expiratory air somewhere in the vocal tract, reducing the volume of the cavity upstream of the closure, so as to create a respiratory depression and then abruptly releasing the closure.

These different processes are called phonation modes. The first three modes require a flow of air which, when expelled from the lungs, passes through the glottis, then into the vocal tract and eventually, into the nasal cavity and out through the lips and nostrils (Figure 4.2). The fourth mode, on the other hand, temporarily blocks the flow of expiratory or inspiratory air. This mode is used to produce “clicks”, implosive consonants present in the phonological system of languages, such as the Xhosa, which is spoken in South Africa. However, clicks are also present in daily non-linguistic production as isolated sounds, with various bilabial, alveodental, pre-palatal and palatal articulations. These sounds are correlated with various meanings in different cultures (kisses, refusal, call, etc.).

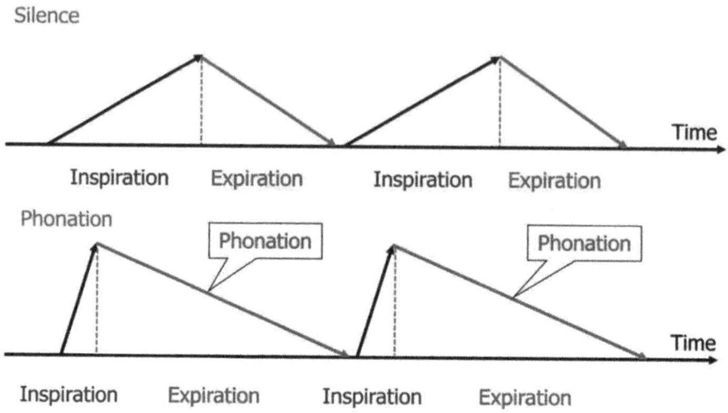

Figure 4.1. *Normal breathing cycle and during phonation. For a color version of this figure, see www.iste.co.uk/martin/speech.zip*

The first three modes of phonation involve a flow of air from the lungs to the lips, and can therefore only take place during the exhalation phase of the breathing cycle. When we are not speaking, the durations of the inspiratory and expiratory phases are approximately the same. The production of speech requires us to change the ratio of inhalation to exhalation durations considerably, so that we have the shortest possible inhalation duration and the longest possible exhalation duration. This means that all of the words that we are intending to utter can be spoken.

This is a complex coping mechanism that takes place during language learning in young children, and is aimed at optimizing the duration of the inhalation phase, to accumulate sufficient air volume in the lungs and ensure the generation of a sequence of sounds in the subsequent production of speech. This planning also involves syntax, in that the inspiratory phase that ends an uttered sequence must be placed in a position that is acceptable from the point of view of syntactic decoding by the listener, since inhalation necessitates silence and therefore a pause. The linguistic code forbids, for example, to place a breathing pause between an article and a name (but does allow a so-called "filled" pause in the form of a hesitation "uh", which can only occur in the exhalation phase).

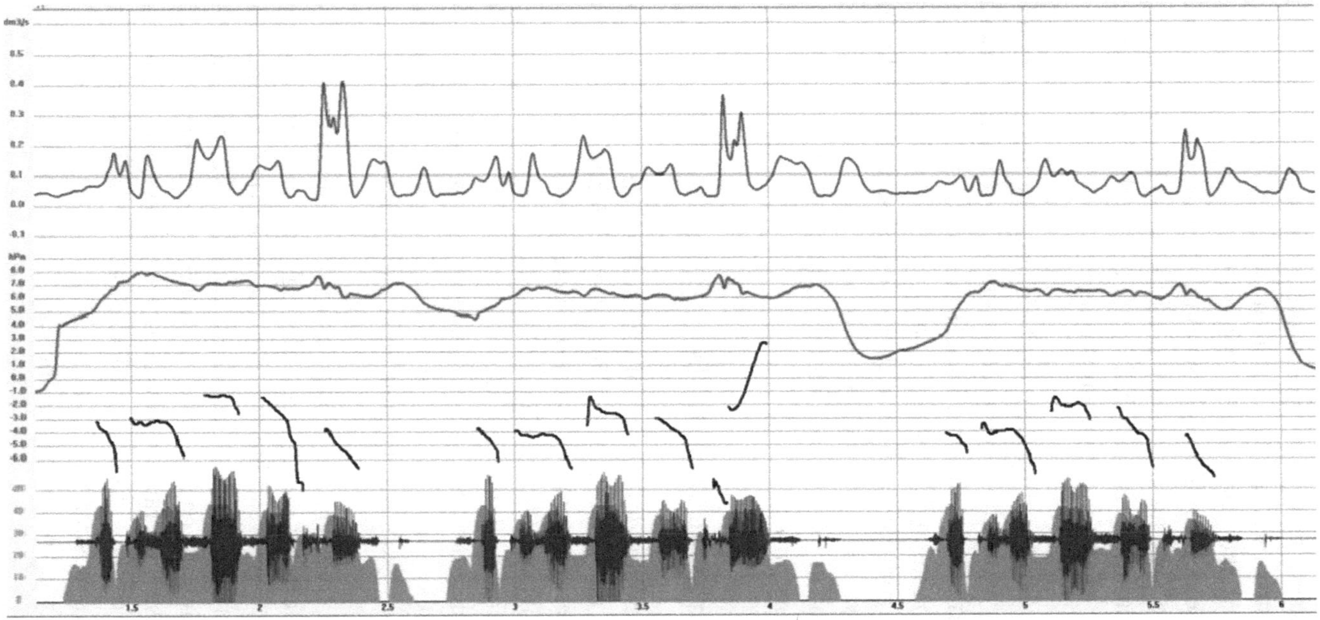

Figure 4.2. *Variation in buccal airflow [dm3/s], subglottic pressure [hPa] and F0 [Hz]: "C'est une chanson triste c'est une chanson triste c'est une chanson triste" (WYH, P01_2, dataD. Demolin). For a color version of this figure, see www.iste.co.uk/martin/speech.zip*

The remarkable feature of speech production lies in the modification of the acoustic structure of the different sources, through changes in the configuration of the vocal tract. Not only can the shape of the duct be modified by the degree of mouth opening, the positioning of the back of the tongue and the spacing or rounding of the lips, but it is also possible to associate the nasal cavity with it through the uvula, which acts as a switch for the passage of air through the nostrils.

The sounds produced by each of the phonatory modes can be "sculpted" in such a way as to allow the production of vowel and consonant sounds, which are sufficiently differentiated from each other by their timbre to constitute a phonological system, all while combining the modes of production. In addition to these rich possibilities, is the ability to modulate the laryngeal vibration frequency. It is as if by speaking, not only is the musical instrument used at each moment changed, but also the musical note played on the instrument.

4.2. Vibration of the vocal folds

A very schematic description of the mechanism of vibration of the vocal folds could be this (there are still impassioned debates on this issue, but we will give the best accepted explanations here, see (Henrich 2001)): the vocal folds (commonly called "vocal cords"), which are in fact two cartilages, are controlled by about 20 muscles, which, to simplify, can be grouped together according to their action when positioning themselves against one another (adductor muscles) and according to the tension applied to them by the tensor muscles, which modify their mass and stiffness.

If the vocal folds are far enough apart, the inspiratory air fills the lungs, and the expiratory air passes freely through the nasal passage, and possibly the vocal tract (by breathing through the mouth). When they are brought closer together and almost put into contact, the constriction produces turbulence during the passage of air (inspiratory or expiratory), which, in the expiratory phase, generates a friction noise (pharyngeal consonants). If they are totally in contact, the flow of expiratory air is stopped, and excess pressure occurs upstream of the vocal folds if the speaker continues to compress the lungs, producing an increase in subglottic pressure.

As soon as there is a sufficient pressure difference between the upstream and downstream sides of the vocal folds that are in contact, and therefore closed, and depending on the supply force, the closure gives way, the vocal folds open (Figure 4.3), and the expiratory air can flow again. Then, an aerodynamic phenomenon occurs (Bernoulli phenomenon, described by the mathematician Daniel Bernoulli, 1700–1782), which produces a depression when the section widens in the movement of the fluid, which is the case for the air as it passes through the vocal folds to reach the pharyngeal cavity. This negative pressure will act on the open vocal folds and cause them to close abruptly, until the cycle starts again.

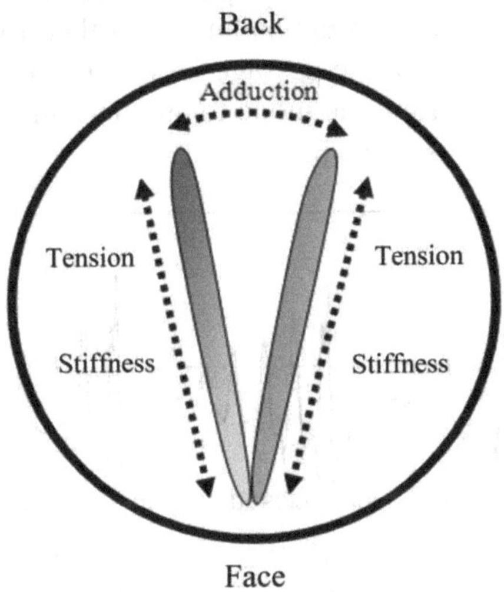

Figure 4.3. *Simplified diagram of the voice fold control system*

The vibration mechanism is therefore controlled by the adductor muscles, which bring the vocal folds into contact with more or less force, and by the tensor muscles, which control their stiffness and tension. The force that the vocal folds are brought into contact with will play an important role in the realization of the opening-closing cycles. When this tension is high, more pressure will be required upstream of the glottis to cause them to open. The closing time will therefore be longer within a cycle. In cases of extreme tension, there will be a "creaky" voice, with irregular opening and closing laryngeal cycle durations, or alternating short and long durations. Conversely, if this tension is too low, and if the adductor muscles do not bring them completely together, the vocal folds will not close completely and air will continue to pass through, despite the vibration (case of incomplete closure). This is known as a "breathy" voice.

The most efficient mode, in terms of the ratio of acoustic energy produced to lung air consumption, occurs when there is a minimal closing time. This mode is also the most efficient when the closure is as fast as possible, producing high amplitude harmonics (Figure 4.4).

The control of the adductor and tensor muscles of the vocal folds allows the frequency of vibration, as well as the quantity of air released during each cycle, to be controlled. This control is not continuous throughout the range of variation and shifts the successive opening and closing mechanisms from one mode to the other, in an abrupt manner. It is therefore difficult to control the laryngeal frequency in these passages continuously over a wide range of frequencies that switch from one mode to another, unless one has undergone specific training, like classical singers.

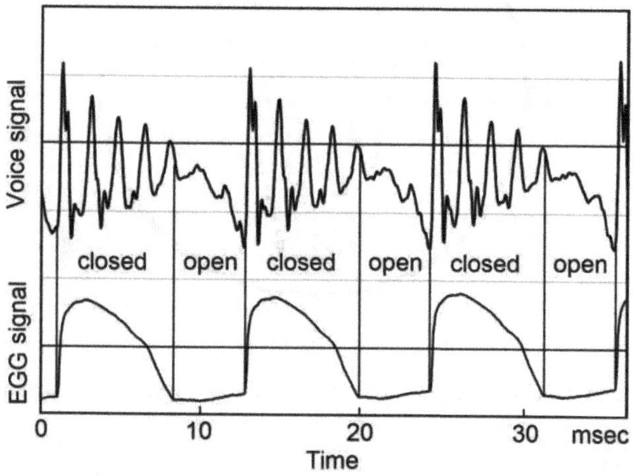

Figure 4.4. *Estimation of glottic waveforms obtained by electroglottography, male speaker, vowel [a] (from (Chen 2016))*

The lowest vibration frequencies are obtained in creaky mode, or vocal fry: the vocal folds are short, very thick and not very tense (Hollien and Michel 1968), and are maintained at the beginning of the cycle, strongly in contact by the adductor muscles. Significant irregularities may occur in the duration of one cycle to the next. In the second, "normal" mode, the vocal folds vibrate over their entire length and with great amplitude. When the frequency is higher, the vibrations only occur over part of the length of the vocal folds, so as to reduce the vibrating mass and thus achieve shorter cycle times. Lastly, in the third mode, called the falsetto or whistle voice, the vocal folds are very tense and are therefore very fine. They vibrate with a low amplitude, producing much less harmonics than in the first two modes.

In the first two modes, creaky and normal, the vibration of the vocal folds produces a spectrum, whose harmonic amplitude decreases by about 6 dB to 12 dB per octave (Figure 4.5). What is remarkable in this mechanism is the production of harmonics due to the shape of the glottic waveform, which has a very short closing time compared to the opening time. This characteristic allows the generation of a large number of vowel and consonant timbres by modifying the relative amplitudes of the harmonics, due to the configuration of the vocal tract. A vibration mode closer to the sinusoid (the case of the falsetto mode) produces few or no harmonics, and would make it difficult to establish a phonological system consisting of sufficiently differentiated sounds, if it were based on this type of vibration alone.

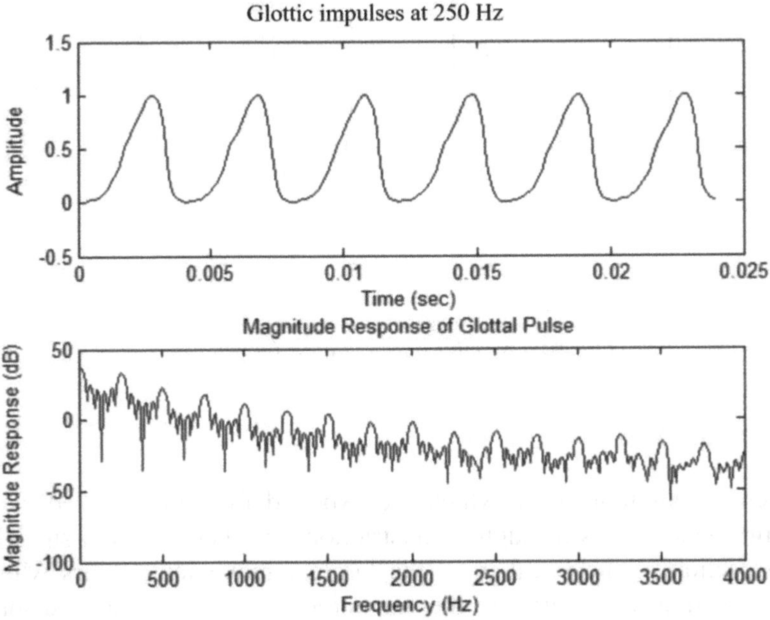

Figure 4.5. *Glottic wave spectrum showing the decay of harmonic peaks from the fundamental frequency to 250 Hz (from Richard Juszkiewicz, Speech Production Using Concatenated Tubes, EEN 540 - Computer Project II)*[1]

1 Qualitybyritch.com: http://www.qualitybyrich.com/een540proj2/.

4.3. Jitter and shimmer

The parameters characterizing the variations in duration and intensity from cycle to cycle are called "jitter" and "shimmer" respectively. The "jitter" corresponds to the percentage of variation in duration from one period to the next: $2\ (t_i - t_{i-1}) / (t_i + t_{i-1})$, and the "shimmer" to the variation in intensity: $2\ (I_{i-1} - I_i) / (I_{i-1} + I_i)$ (Figure 4.6). The statistical distribution of these parameters is characterized by a mean and a standard deviation, reflecting the dispersion of these values around the mean. In speech-language pathology, the standard deviation is indicative, as is the symmetry or asymmetry of the distribution, of certain physiological conditions affecting the vocal folds.

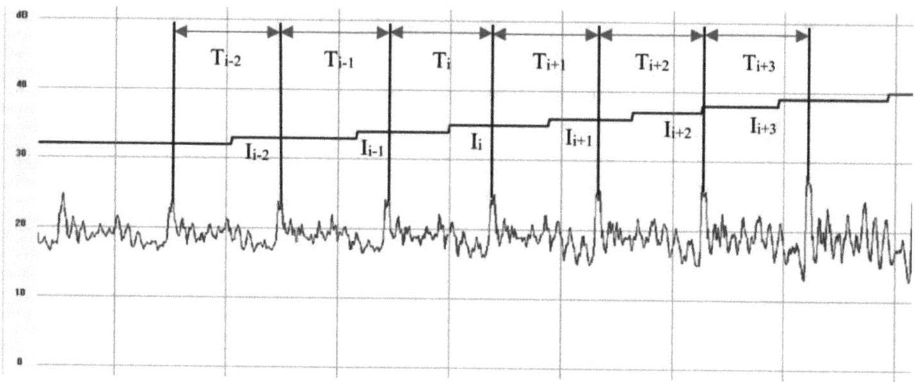

Figure 4.6. *Jitter and shimmer (vowel (a))*

4.4. Friction noises

When the air molecules, which are expelled from the lungs during the exhalation phase, pass through a constriction, in other words, a sufficiently narrow section of the vocal tract, the laminar movement occurs when the passage section is disturbed: the air molecules collide in a disorderly manner, and produce noise and heat, in addition to the acceleration of their movement. It is this production of noise, comprising *a priori* all the components of the spectrum (similar to "white noise", for which all the spectral components have the same amplitude, just as the white light present in the analysis has all the colors of the rainbow), which is used to produce the frictional consonants. The configuration of the vocal tract upstream and downstream of the constriction, as well as the position of the constriction in the vocal tract, also allows the amplitude distribution of the frictional noise components to be modified from approximately 1,000 Hz to 8,000 Hz.

A constriction that occurs when the lower lip is in contact with the teeth makes it possible to generate the consonant [f], a constriction between the tip of the tongue and the hard palate generates the consonant [s], and between the back of the tongue and the back of the palate generates the consonant [ʃ] (ʃ as in *short*). There is also a constriction produced by the contact of the upper incisors to the tip of the tongue for the consonant [θ] (θ as in *think*).

4.5. Explosion noises

Explosion noises are produced by occlusive consonants, so called because generating them requires the closure (occlusion) of the vocal tract so that excess pressure can be created upstream of the closure. This excess pressure causes an explosion noise when the closure is quickly released and the air molecules move rapidly to equalize the pressure upstream and downstream. These consonants were called "explosive" in the early days of articulatory phonetics. The location of the closure of the vocal tract, called the place of articulation, determines the acoustic characteristics of the signal produced, which are used to differentiate the different consonants of a phonological system in hearing.

In reality, these acoustic differences are relatively small, and it is more the acoustic effects of the articulatory transitions, which are necessary for the production of a possible vowel succeeding the occlusive, that are used by listeners. In this case, the vibrations of the vocal folds, shortly after the occlusion is relaxed (the so-called voice onset time or VOT, specifically studied in many languages), cause the generation of a pseudo vowel with transient spectral characteristics, that stabilize during the final articulation of the vowel. A transition of formants (see Chapter 6) occurs, in other words, resonant frequencies determined by the configuration of the vocal tract, which are used by the listener to identify the occlusive consonant, much more than the characteristics of the explosion noise. Nevertheless, it is possible to recognize occlusive consonants pronounced in isolation in an experimental context, such as for the occlusive consonants [p], [t] and [k], which are produced respectively by the closure of the lips (bilabial occlusive), the tip of the tongue against the alveoli of the upper teeth (alveolar), and the back of the tongue against the hard palate (velar).

4.6. Nasals

Nasal vowels and consonants are characterized by the sharing of the nasal passage with the vocal tract, by means of the uvula, which acts as a switch. This additional cavity inserted in the first third of the path of exhaled air, and modulated by the vocal folds (or by turbulent air in whispered speech), causes a change in the source's harmonic resonance system, which can be accounted for by a mathematical model (see Chapter 8). The appearance of nasal vowel bandwidth formants, that are larger than with the corresponding oral vowels (which is difficult to explain by observation of their spectral characteristics), is thus easily elucidated by this model. In French, the nasal vowels used in the phonological system are [ã], [õ] and [ɛ̃], and the nasal consonants [m], [n], [ɲ] as in *agneau* and [ŋ] as in *parking*.

4.7. Mixed modes

The apparatus of phonation can implement the vibrational voicing of the vocal folds simultaneously with the other modes of phonation, thus opposing vowels or voiced consonants to their articulatory correspondents known as being voiceless, without vibrating the vocal folds. Thus [v] is generated with a friction noise and the vibration of the vocal folds, but with an articulatory configuration close to [f]. It is the same between [s] and [z], [ʃ] and [ʒ] (the phonetic symbol ʃ, as in *short* and the symbol ʒ, as in *measure*). In this mixed mode, the [v], [z] and [ʒ], which must vibrate at the same time and allow enough air to pass through to allow friction noise, produce harmonics of much less amplitude than vowels.

4.8. Whisper

It is possible to generate vowels and consonants without vibrating the vocal folds, with only friction noise. In the case of vowels, the source of friction is located at the glottis and is produced by a tightening of the vocal folds sufficient to produce enough turbulence in the airflow. The final intensity produced is much lower than the neighboring vowel production mode, which the speaker can compensate for by, for example, making the vowels accentuated, by increasing their duration compared to their normal duration.

4.9. Source-filter model

In order to represent all these mechanisms in a simplified manner, a speech production model called a source-filter is often used (Figure 4.7): the source consists of a train (a sequence) of pulses of frequency F0 (reciprocal of the time interval between each pulse) and a noise source, the amplitudes of which are controlled by a parameter A. The F0 frequency, (also called fundamental frequency, bringing an unfortunate confusion with the Fourier analysis fundamental frequency), corresponds to the laryngeal frequency, and the noise source to the friction noise in the phonation. A mathematical model of the vocal tract incorporates the spectral characteristics of the glottic source and the nasal tract. This model also incorporates an additional filter that accounts for the radiation characteristics at the lips. This type of model accounts fairly well for the (approximate) independence of the source from the vocal tract configuration.

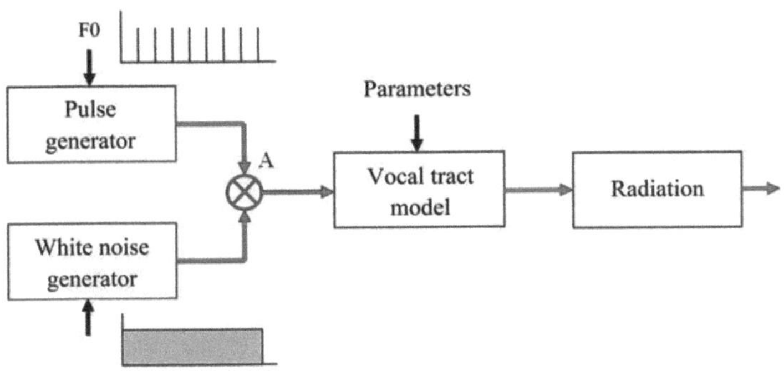

Figure 4.7. *Speech production model*

Acoustic descriptions of speech sounds make extensive use of this model, which separates the source of the sound and the sculpture of its harmonic spectrum through the vocal and nasal passages very well. It helps to understand that speech characteristics such as intonation, due to variations in laryngeal frequency over time, are independent of the timbre of the sounds emitted, at least as a first approximation.

5

Source-filter Model Analysis

5.1. Prony's method – LPC

Acoustic analysis of the speech signal by Fourier harmonic series is, by design, completely independent of the nature of the signal, and its results are interpretable for sound production by humans, as well as by chimpanzees or sperm whales. Although its most important limitation is on transient phenomena, which are not inherently periodic, the representation of the analysis by an amplitude-frequency spectrum corresponds quite well to the perceptive properties of human hearing. On the other hand, the so-called Prony method (Gaspard François Clair Marie, Baron Riche de Prony 1755–1839) is very different in that the principle of analysis involves a model of phonation, which *a priori* makes this method unsuitable for the acoustic analysis of sounds other than those produced by a human speaker.

The Fourier and Prony methods also differ in principle by an essential property. Fourier analysis generates a harmonic representation, i.e., all components are pure sounds whose frequencies are integer multiples of a basic frequency called the fundamental frequency. Prony analysis generates a non-harmonic representation, in other words, all components are damped pure sounds whose frequencies are not integer multiples of a fundamental frequency.

In speech analysis, Prony's method, also known as the Linear Prediction Coefficients (LPC) method, is a generic term for solving equations describing a source-filter model of phonation from a segment of speech signal. It is therefore, in principle, very different from Fourier-series analysis. Instead of proposing amplitude spectra obtained by harmonic analysis, Prony's analysis involves a source-filter model, whose parameters characterizing the filter (representing the vocal tract) are adjusted so that, when stimulated by an impulse train whose period corresponds to the laryngeal frequency for a given time window, or by white noise simulating a source of friction, the filter solicited produces a signal at the output, one as close as possible to a segment of the original signal (Figure 5.1).

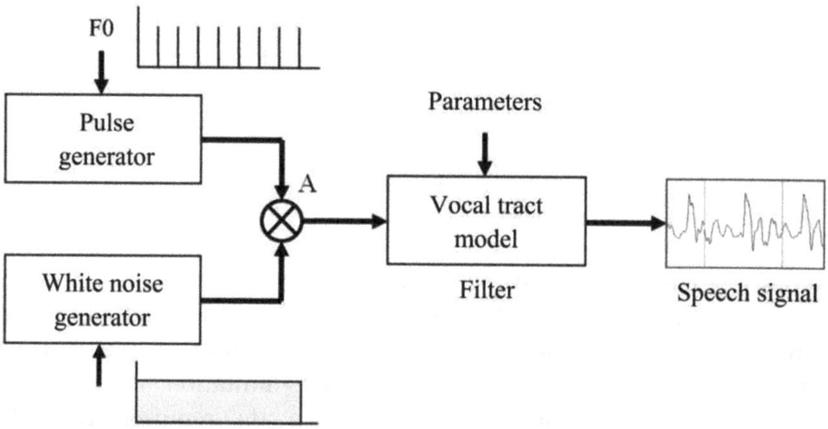

Figure 5.1. *Implicit model in LPC analysis*

Prony's method thus proceeds according to an implicit model, i.e., a symbolic construction that more or less simulates the reality of the phonatory mechanism. Thus, within an analyzed speech segment, the more or less regular cycles of laryngeal impulses are represented by a train of impulses of constant frequency, impulses which produce a large number of harmonics of constant amplitude, whereas in reality, they have an amplitude decreasing from 6 dB to 12 dB per octave. Furthermore, the friction noise source is positioned at the same place as the pulse source in the model, which does not correspond to reality, except for the laryngeal consonant [h]. In fact, the position of the friction source in the vocal tract for the fricative consonants [f], [s] and [ʃ] is, respectively, at the lips, at the back of the alveoli of the teeth of the upper jaw and at the top of the hard palate.

The interest of such a model essentially lies in the possibility of directly obtaining the resonance frequencies of the filter, the model of the vocal tract, and thus of estimating the formants – areas of reinforced harmonics – without having to make a visual or algorithmic interpretation of a spectrum or spectrogram, which is not always obvious. This is due to the fact that in this method, the data – i.e., the windowed segments of the signal – are forced to correspond to the output of the source-filter model. The formants obtained actually correspond to the frequencies of the maxima of the filter response curve. The adequacy of the overall response curve of the vocal tract, with the maxima that produced the analyzed signal, is not necessarily guaranteed.

5.1.1. *Zeros and poles*

Electrical filters can generally be defined by a mathematical equation called a transfer function that accounts for the response of the filter (the output) to a given stimulation (an input). For some types of filters, transfer functions can be expressed as a fraction, with the numerator and denominator being polynomial functions of frequency (a polynomial function of a variable is a sum of terms equal to a coefficient multiplied by a power of the variable). It is then possible to calculate the frequency and phase response of the filter from a transfer function. However, the polynomial functions of the numerator and denominator may have particular frequency values that cancel them out, making the transfer function zero for the numerator and infinite for the denominator (unless the same frequency simultaneously makes the numerator and denominator zero). When a frequency cancels out the numerator, the transfer function is referred to as zero, and when it cancels out the denominator, it is referred to as a pole. The amplitude response curve characterizing the filter therefore has zero values for zeroes of the transfer function, and infinite values for poles. In other words, nothing exits the filter for a frequency value resulting in a zero for the numerator of the transfer function, and an infinitely large signal exits the filter for a frequency value resulting in a zero for the denominator of the transfer function.

The interest of the transfer functions for speech analysis comes from the connection that can be made between the speech sound generation mechanism (particularly for vowels) and the source-filter model: the successive laryngeal cycles are represented by a pulse train (a sequence of impulses with a period that is equal to the estimated laryngeal period), and the frictional noise of the fricatives by white noise (white noise comprises all frequencies of equal amplitude in the spectrum).

The source-filter model is thus an approximation of reality, insofar as, on the one hand, the glottal stimulation is not an impulse train, and on the other, the source of fricative sounds is not positioned in the same place in the vocal tract. Likewise, the spectrum of the source of laryngeal vibration, characterized by a drop from 6 dB to 12 dB, can be taken into account by integrating a very simple single-pole filter into the vocal tract model in the transfer function. For the rest, if the filter has appropriate characteristics, the poles should correspond to formants, which are effectively values of the laryngeal frequency that correspond to a reinforcement of the amplitudes of the harmonics of the laryngeal frequency.

The principle of Prony's analysis, and therefore of the calculation of linear prediction coefficients, is to determine the coefficients of a suitable filter that models the characteristics of the vocal tract (by integrating the characteristics of the source). Since the mathematical formulation of the problem involves stationarity, it will be necessary, like in Fourier harmonic analysis, to take windows of a sufficient minimum duration from the signal to solve the system of equations and of a maximum acceptable duration with respect to the stationarity of the vocal tract. The minimum duration is a function of the number of signal samples required, therefore also of the sampling frequency, and also of the polynomial degree p of the equation describing the filter.

We then put the following equation forward:

$$x_n = \sum_{m=1}^{p} a_p x_{n-m}$$

which simply means that the signal value at time n (these are sampled and indexed values 0, 1, ..., n) results from the sum of the products of the signal values at times $n-1$, $n-2$, ..., $n-p$. We can show, (not here!) by calculating the z-transform (equivalent to the Laplace transform for discrete systems, i.e., the sampled values), that this equation describes an autoregressive type of filter (with a numerator equal to 1), which, to us, should correspond to a model of the vocal tract valid for a small section of the signal, that is, for the duration of the time window used. The mathematical description of this filter will be obtained when we know the number of m values and what these values are for the m coefficients a_p. The transfer equation of the corresponding all-pole model is:

$$H(z) = \frac{B(z)}{A(z)} = \frac{1}{a_1 + a_2 z^{-1} + \cdots + a_{m-1} z^{-m}}$$

which is the equation for an autoregressive model, abbreviated to model AR. To obtain the values of these coefficients, the prediction given by this equation, and hence the output of the filter, is compared with the reality. In other words, compared to a certain number of successive samples of the signal, by minimizing the difference between the prediction and the reality of the signal, for example by the method of least squares. Mathematically, it will therefore be a matter of minimizing the error ε defined by $\varepsilon = \sum_{n=1}^{k} \sqrt{(x_n - s_n)^2}$ where x_n = predicted signal samples and s_n = signal samples. It is conceivable that the filter obtained will be all the more satisfactory as the prediction coefficients will minimize the error over a sufficient duration. However, this error will reach a maximum when the underlying model is no longer valid, that is, predominantly at the time of the laryngeal pulses. These prediction error maxima for voiced sounds are called prediction residuals.

5.2. Which LPC settings should be chosen?

5.2.1. *Window duration?*

Minimizing prediction error means solving a system of linear equations with unknown p, where p is the number of coefficients and corresponds to the order of the filter. The minimum necessary number of samples k is equal to p, so for a sampling frequency of 16,000 Hz, and a filter of the order of 12, we need a time sampling window duration of 12/16,000 = 0.75 ms, which is much less than that necessary for the measurement of formants by Fourier harmonic analysis! However, in this case, the estimation of the formants from the poles of the transfer function will only be valid for a very limited duration of the signal. It is therefore preferable to choose a longer time window and obtain the optimized linear prediction coefficients over a duration that is supposed to be more representative of the signal. Figure 5.2 shows that although the 16 ms and 46 ms windows give almost identical spectra, the 2 ms spectrum is still relatively usable. The advantage of a relatively large time window lies in the calculation of the prediction error, which, when carried out on a larger number of signal samples, gives a more satisfactory approximation.

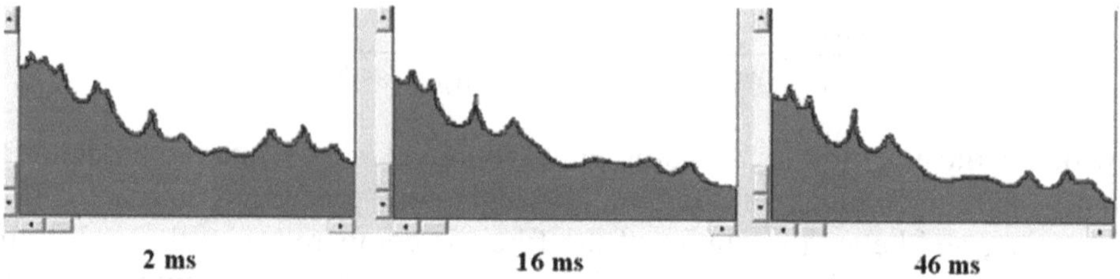

Figure 5.2. *Comparison of Prony spectra of the order of 12 with a window of 2 ms, 16 ms and 46 ms*

Different methods exist to solve this system of equations, known among others as the correlation method, the covariance method and Burg's method. The latter method is the most widely used today as it guarantees stable results with a reasonable calculation time.

5.2.2. *What order for LPC?*

The number of prediction coefficients determines the number of poles and thus the number of peaks in the system response curve. A heuristic rule specifies that there is generally 1 formant (thus 2 poles) per kHz of bandwidth. Two poles are added to this value to account for the spectral characteristics of the glottic source. For the value of a 16,000 Hz sampling frequency, we thus obtain an order p = 18 or 9 poles, at 22,050 Hz, p = 24 or 12 poles, which are satisfactory approximations in practice.

5.3. Linear prediction and Prony's method: nasals

The calculation of the filter resonance frequencies of a source-filter model (whose source is a pulse train) amounts to representing the sampled signal by a sum of damped sinusoids (responding to each of the input pulses of the filter), whose frequencies are equal to the resonance frequencies of the filter, which corresponds to the definition of Prony's method.

The source-filter model for the nasal vowels must have an additional element that takes into account the communication of the nasal cavities with the vocal tract at the uvula level. It can be shown that the all-pole model (ARMA) is no longer valid and that an equation for the transfer function with a non-zero numerator should be considered:

$$H(z) = \frac{B(z)}{A(z)} = \frac{b_1 + b_2 z^{-1} + \cdots + b_{n+1} z^{-n}}{a_1 + a_2 z^{-1} + \cdots + a_{m-1} z^{-m}}$$

The values of z that cancel out the numerator are called the filter zeros, those that cancel out the denominator are called the poles. The existence of zeros in the response curve can be found in the calculation of articulatory models involving nasals (see Chapter 8). This model is an ARMA model, an acronym for Auto Regressive Moving Average.

Solving this equation uses a variation of the covariance method to solve the all-pole equation, in order to determine the denominator coefficients of the transfer function, and then calculates the numerator coefficients so that the impulse response of the model filter exactly matches the first n+1 samples of the signal (Calliope 1989).

5.4. Synthesis and coding by linear prediction

Linear prediction coefficients define the characteristics of a model vocal tract filter, optimized for a certain duration of the speech signal, and, in any case, for the duration of a laryngeal period. After storing them, by restoring these coefficients from segment to segment, it is then possible to generate a speech signal whose source parameters are controlled independently of the characteristics of the vocal tract, simulated by the linear filter. A prosodic morphing process is thus carried out, a speech whose formant characteristics are preserved and whose source factors, laryngeal frequency, intensity and duration, are manipulated.

If the analysis is carried out under conditions which sufficiently exclude any noise in the speech signal, and the source and the laryngeal frequency in particular is of good quality, a very efficient coding system is available, given that for the duration of one or more laryngeal cycles, the sampled values of the signal are coded by p prediction coefficients (in addition to the coding of the source, the period of the laryngeal pulse, its amplitude and that of the source of any friction noise). Thus, for a speech signal sampled at 16,000 Hz, and an analyzed segment of 30 ms, the 480 samples of the segment can, for example, be coded by 12 parameters in addition to the source values. This technique was first popularized by the educational game *Speak and Spell* in the 1980s, where a relatively large number of words were orally synthesized from a very limited LPC coefficient memory.

The LPC method was discovered by researchers at the Bell Laboratories in Murray Hill and involved geodetic analysis for earthquake prediction. In 1975, the adaptation of LPC analysis for speech was published, but it was mainly the applications in synthesis that were highlighted (in the issue of JASA, the American journal of acoustics presenting this method, the authors even inserted a floppy disc that allowed the examples of synthesis to be listened to and their quality to be appreciated). They thus demonstrated the extraordinary advantages of a synthesis whose source parameters could be very easily manipulated (noise source for fricatives, pulse source of variable period for vowels and voiced consonants), and ensured the adequacy of the filter representing the vocal tract through the analysis of successive windows of the signal (and also incorporating the spectral parameters of the source). The authors were also careful to avoid the presence of stop consonants in their examples, which were all similar to the "all lions are roaring" sentence, with no stop consonants that were not directly taken into account by the source-filter model.

Undoubtedly offended by the resonance of the LPC method and the promotion provided by the communication services of Bell (installed on the East Coast of the United States), the researchers of the West Coast endeavored to demonstrate that the method of resolution of the filter that was modeling the vocal tract was only taking up the method of resolution of a system of equations that had been proposed by Prony in 1792, and had fun each time quoting the *Journal de l'École polytechnique* in French in the references to the various implementations that they published.

Today, experts in speech signal analysis, inheritors of early research, still refer to the LPC method. Most of them are signal processing engineers who are more concerned with speech coding for telephone or Internet transmission. On the other hand, researchers who are more interested in phonetic research and voice characterization refer to Prony's method.

Box 5.1. *The rediscovery of Prony. East Coast versus West Coast*

6

Spectrograms

6.1. Production of spectrograms

The spectrogram, together with the melody analyzer (Chapter 7), is the preferred tool of phoneticians for the acoustic analysis of speech. This graphical representation of sound is made in the same way as cinema films, by taking “snapshots” from the sound continuum, analyzed by Fourier transform or Fourier series.

Each snapshot results in the production of a spectrum showing the distribution of the amplitudes of the different harmonic components; in other words, a two-dimensional graph with frequency on the x-axis and amplitude on the y-axis. To display the spectral evolution over time, it is therefore necessary to calculate and display the different spectra on the time axis through a representation of time on the abscissa, frequency on the ordinate, and amplitude as a third dimension coded by color or a level of gray.

Theoretically, considering the speed of changes in the articulatory organs, the necessary number of spectra per second is in the order of 25 to 30. However, a common practice is to relate the number of temporal snapshots to the duration of the sampling window, which in turn determines the frequency resolution of the successive spectra obtained. This is done by overlapping the second temporal half of a window with the next window.

We have seen that the frequency resolution, i.e., the interval between two values on the frequency axis, is equal to the reverse of the window duration. In order to be able to observe harmonics of a male voice at

For a color version of all the figures in this chapter, see: www.iste.co.uk/martin/speech.zip.

100 Hz, for example, a frequency resolution of at least 25 Hz is required, that is, a window of 40 ms. A duration of 11 ms, corresponding to a frequency resolution of about 300 Hz, leads to analysis snapshots every 5.5 ms. A better frequency resolution is obtained through a window of 46 ms, which corresponds to the overlap with the spectrum every 23 ms (Figure 6.1). Spectrograms implemented in programs such as WinPitch perform the graphical interpolation of the spectrum intensity peaks, so as to have a global graphical representation that does not depend on the choice of window duration.

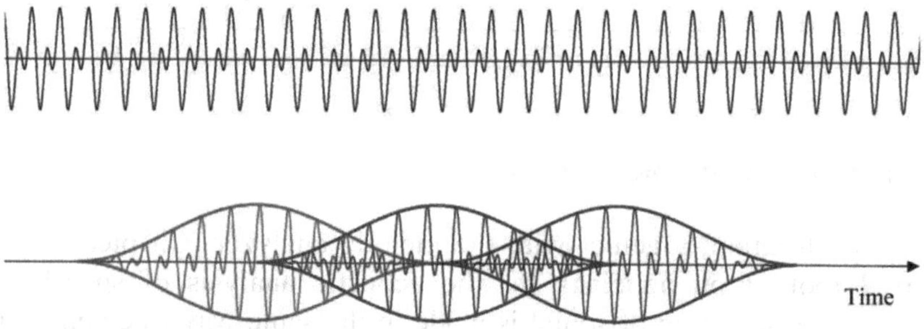

Figure 6.1. *Overlapping analysis windows*

Graphic interpolation can also be performed on the time axis. However, with the execution speed of today's computer processors, rather than interpolation, the simplest way is to drag the time sampling window to match the number of spectra to the number of pixels on the graphics display (or possibly the printer), regardless of the duration of the signal displayed on the screen. The result is a highly detailed spectrographic image in both the time and frequency axes. In reality, the information available on the frequency axis results from the properties of Fourier analysis, as seen in Chapter 5, which gives n/2 frequency values for a speech segment represented by n samples. The appearance of a continuous spectrum on the frequency axis results from interpolation (which can be sub-interpolation if the number of pixels in the display is less than the number of frequency values).

A spectrogram makes a three-dimensional representation: time on the horizontal axis, the frequency of the harmonic (for Fourier) or non-harmonic (for Prony) components on the vertical axis, and the intensity of the different components on an axis perpendicular to the first two axes is encoded by the level of gray (or color coding, but these are not very popular with phoneticians). On a computer screen or on paper, the time, frequency and intensity of each component are interpolated so as to appear in the traditional appearance of analogue spectrograms of the 1960s.

We have seen above that the duration of the time window, defining the speech segment to be analyzed at a given instant, determines the frequency resolution of the successive spectra represented on the spectrogram. The first observation to make is that a constant frequency sound is represented by a horizontal line on the spectrogram, the thickness of which depends on the type and duration of the window used (most speech spectrography software uses a default Hann(ing) window). Figure 6.2 shows some examples of analysis of a sound at 1000 Hz constant frequency.

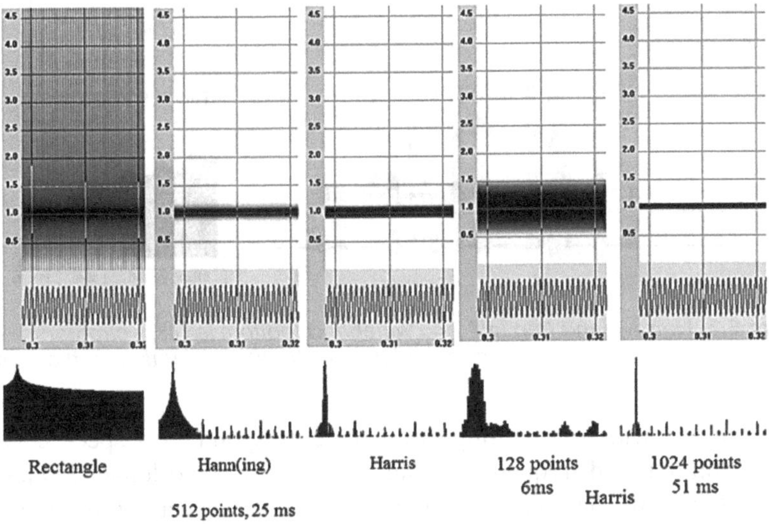

Figure 6.2. *Pure sound analyzed at 1,000 Hz, respectively, with a 25 ms rectangular window, a 25 ms Hann(ing) window, a 6 ms Harris window and a 51 ms Harris window*

It can be seen that the Harris window (although little used in acoustical phonetics) gives the best results for the same window duration.

Figure 6.3 shows two examples of frequency interpolation: while the spectra at a given instant (on the right of the figure) show stepped variations (narrowband at the top of the figure and wideband at the bottom), the corresponding spectrograms (on the left of the figure) show a continuous aspect in both the time and frequency axes. It can also be seen that the wideband setting blurs the harmonics, so that the areas with formants are more apparent.

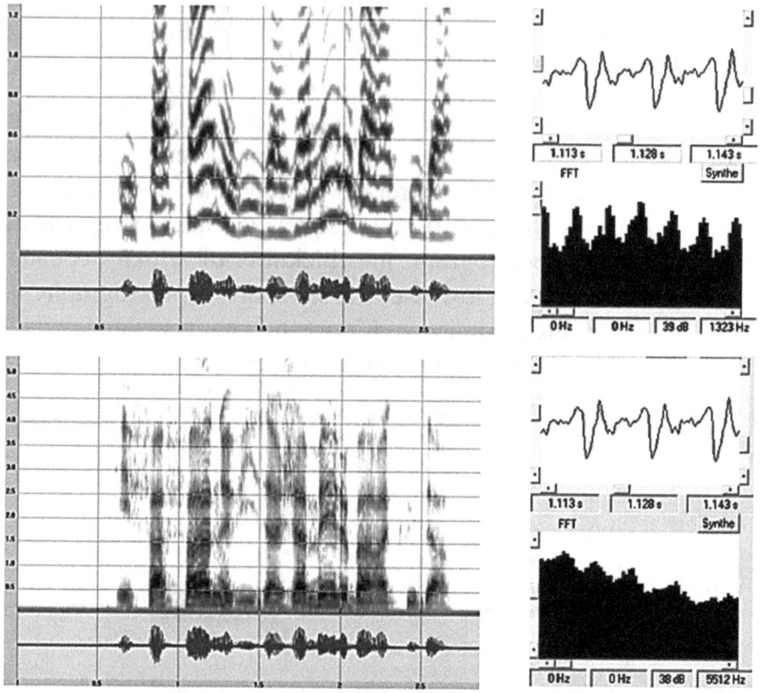

Figure 6.3. *Narrowband (top left) and wideband (bottom left) spectrograms. To the right of the figure are the corresponding spectra at a given time of the signal*

In fact, the wideband setting does not correspond to a specific value of window duration and frequency resolution. Since it is no longer a question of distinguishing one harmonic from another in the spectrogram, the correct wideband setting depends on the spacing of the harmonics, and thus on the laryngeal frequency of the voice. A window duration value of 16 ms will be adequate for a male voice, but will not necessarily be adequate for a female voice, which requires an 8 ms or 4 ms window to obtain a wideband spectrogram.

6.2. Segmentation

6.2.1. *Segmentation: an awkward problem (phones, phonemes, syllables, stress groups)*

Contemporary linguistic theories (i.e., structural, generative-transformational) operate on units belonging to several levels, some of which are very much inspired by, if not derived from, the alphabetical writing system, predominantly of English and French, including punctuation. Thus, a sentence is defined as the space between two points, the word as the unit between two graphic spaces, and the phoneme as the smallest pertinent sound unit. The phoneme is the unit revealed by the so-called minimal pairs test operating on two words with different meanings, for example *pier* and *beer*, to establish the existence of |p| and |b| as minimum units of language. This test shows that there are 16 vowels in French, and about 12 vowels and 9 diphthongs in British English. However, a non-specialist who has spent years learning to write in French will spontaneously state that there are 5 vowels in French (from the Latin): a, e, i, o and u (to which "y" is added for good measure), whereas the same non-specialist will have no problem identifying and quoting French syllables.

6.2.2. *Segmentation by listeners*

In reality, these units – phoneme, word or sentence – are not really the units of speakers and listeners, at least after their period of native language acquisition. Recent research on the links between brain waves and speech perception shows that the units operated by speaking subjects do not operate with phonemes but with syllables as minimal units. In the same way, speakers and listeners use stress groups (or rhythmic groups for tonal languages) and not spelling words for syllable groups; they use prosodic structures to constitute utterances, and lastly breath groups bound by the inhalation phases of the breathing cycles that are obviously necessary for the survival of speaking subjects.

Indeed, we do not read or produce speech phoneme after phoneme (or phone after phone, the phone being the acoustic counterpart of the abstract phoneme entity), nor do we read or produce a text word by word (except when acquiring a linguistic system, or when faced with unknown words, in which case we fall back on a process operating syllable by syllable, or even phone by phone). Segmentation by phoneme and by word is thus the result of a transfer of formal knowledge, derived from writing and imposed on phoneticians and speech signal processing specialists by most linguistic theories. In the same way, the sentence is not the space between two points, which is obviously a circular definition referring to the written word, but a sequence of speech ending with specific melodic or rhythmic movements that indicate its limits.

Each of these speech units, as used by speaking subjects, is indicated by a specific device (Martin 2018b):

– syllables: synchronous with theta brain oscillations, varying from 100 ms to about 250 ms (i.e. from 4 Hz to 10 Hz);

– stress groups: characterized by the presence of a single non-emphatic stressed syllable, synchronized with delta brain oscillations, ranging from 250 ms to about 1,250 ms (i.e. from 0.8 Hz to 4 Hz);

– breath groups: delimited by the inhalation phase of the speaker's breathing cycle, in the range of 2 to 3 seconds.

A further difficulty arises from the gap between the abstract concept of phoneme and the physical reality of the corresponding phones, which can manifest itself in various ways in terms of acoustic realization. For example, the same phoneme |a| may appear in the speech signal with acoustic details, such as formant frequencies, varying from one linguistic region to another. However, the major problem is that the boundaries between phones, and even between words, are not precisely defined on the time axis, their acoustic production resulting from a set of gestures that are essentially continuous in time. Indeed, how can you precisely determine the end of an articulatory gesture producing an [m] to begin the generation of an [a] in the word *ma* when it is a complex sequence of articulatory movements? This is, however, what is tacitly required for the segmentation of the speech signal.

Therefore, it seems *a priori* futile to develop speech segmentation algorithms that would determine the exact boundaries of phones (or worse conceptually, phonemes) or words (except, perhaps, preceded or followed by silence). It seems more acceptable to segment, not by determining the "left" or "right" boundaries of the units in question, (i.e. their beginning and end) but rather by a relevant internal position: for example, the peak vowel intensity for the syllable, and the peak stressed vowel intensity for the stress group. The boundaries of these units are then determined in the eventual orthographic transcription by linguistic properties, syllable structure and the composition of the stress groups of the language.

6.2.3. *Traditional manual (visual) segmentation*

The fact remains that the doxa operating on implicitly graphically-based units imposes a segmentation into phones, words and sentences. In practice, the only concession made in relation to writing is to use phonetic notation (IPA or a variant such as SAMPA, Speech Assessment Methods Phonetic Alphabet) rather than their orthographic representation, which is obviously highly variable in French or English for the same vowel, but which is conceivable for languages such as Italian or Spanish.

With these statements in mind, after having obtained a wideband spectrogram more suitable for visual segmentation, the first operation consists of making a phonetic transcription of it, preferably using the characters defined in the IPA. Table 6.1 lists the symbols used for French and English, with corresponding examples.

To illustrate the different practical steps of segmentation on a spectrogram, we will use two short recordings whose spelling transcriptions are "*et Fafa ne visa jamais le barracuda*" (example pronounced by G.B.) for French and "She absolutely refuses to go out alone at night" (from the Anglish corpus) for English.

French

IPA	Consonants
[b]	bien
[p]	pierre
[d]	dent
[t]	temps
[g]	garçon
[k]	quand
[v]	vent
[f]	feuille
[z]	zéro
[s]	sale
[ʒ]	jatte
[ʃ]	chatte
[n]	non
[m]	mur
[ɲ]	agneau
[ŋ]	parking
[l]	lame
[R]	rendre
[r]	variant of [R]

IPA	Vowels
[i]	petit
[e]	chanté
[ɛ]	pelle
[a]	ami
[ɔ]	sol
[u]	bout
[y]	plus
[œ]	meuble
[ə]	le
[ø]	vœux
[o]	eau, saule
[ɛ̃]	main
[ɑ̃]	jean
[ɔ̃]	son
[œ̃]	brun

IPA	Semi-consonants
[j]	yeux
[w]	oui
[ɥ]	huit

English

IPA	Consonants
[p]	Peter
[t]	Tony
[k]	Christopher
[b]	Bob
[d]	David
[g]	Gary
[f]	fool
[θ]	thing
[s]	sing
[ʃ]	short
[tʃ]	China
[v]	van
[ð]	mother
[z]	zone
[ʒ]	measure
[dʒ]	John
[r]	Carry
[l]	Love
[m]	Mother
[n]	None
[ŋ]	king

IPA	Vowels
[æ]	cat
[e]	bed
[ɪ]	sick
[i:]	feel
[a:]	car
[ɒ]	lock
[ɔ:]	ball
[ʊ]	book
[u:]	food
[ə]	of
[ə:]	bird
[ʌ]	Love

IPA	Semi-consonants
[j]	yes
[w]	water

IPA	Diphthongs
[ai]	fly
[ei]	May
[oi]	boy
[aʊ]	now
[iʊ]	fill
[oʊ]	rose
[iə]	beer
[ɛə]	bear
[uə]	poor

Table 6.1. *Phonetic symbols for French and English*

6.2.4. *Phonetic transcription*

The first step consists of a narrow phonetic transcription ("narrow" meaning detailed) of the sound realization: [efafanəvizaʒamɛləbarakyda] for *et Fafa ne visa jamais le barracuda* and [ʃi:æbsIlu:tlɪrɪfju:zɪstʊgoʊʊaIʊtIl oʊʊnætnaIi:t] for *She absolutely refuses to go out alone at night.*

6.2.5. *Silences and pauses*

The second operation consists of identifying possible pauses and silences, for which only the background noise spectra possibly appear on the spectrogram, usually in the form of a horizontal bar at around 100 Hz (Figures 6.4 and 6.5).

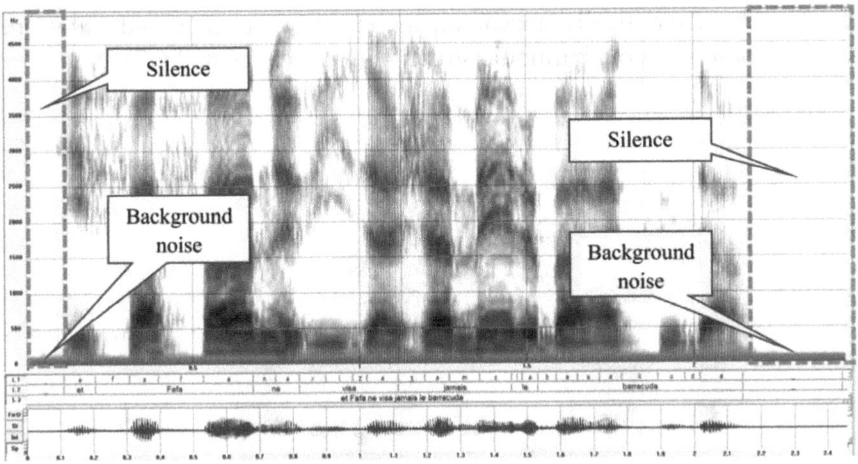

Figure 6.4. *Locating silences in [efafanəvizaʒamɛləbarakyda] (phonetic transcription* of “et Fafa ne visa jamais le barracuda”*)*

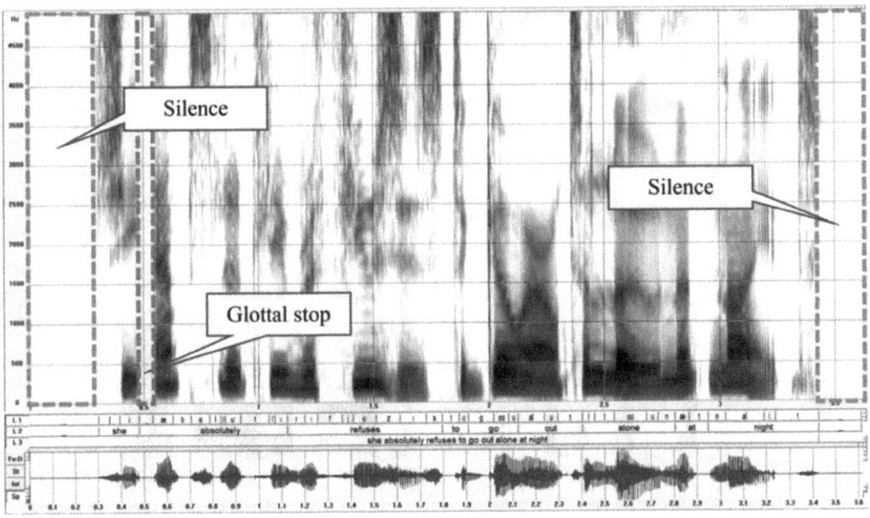

Figure 6.5. *Locating silences in [ʃi:æbsllu:tlɪrɪfju:zɪztʊgoʊʊalʊtlloʊʊnætnali:t] (phonetic transcription of “She absolutely refuses to go out alone at night”)*

6.2.6. *Fricatives*

After identifying the fricative consonants in the phonetic transcription, they must then be identified in the spectrogram. Whether voiced or unvoiced, the fricatives characteristically appear on the spectrogram as clouds of more or less dark points, without harmonic structure (Figures 6.6 and 6.7).

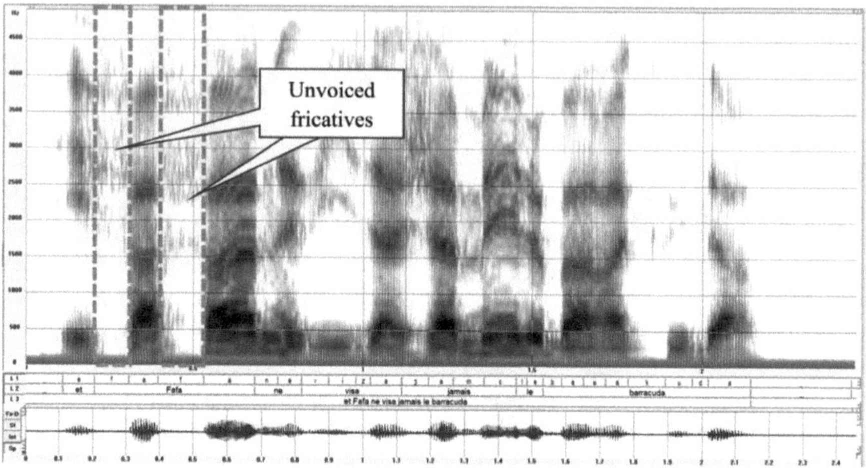

Figure 6.6. *Locating unvoiced fricatives [efafanəvizaʒamɛləbarakyda]*

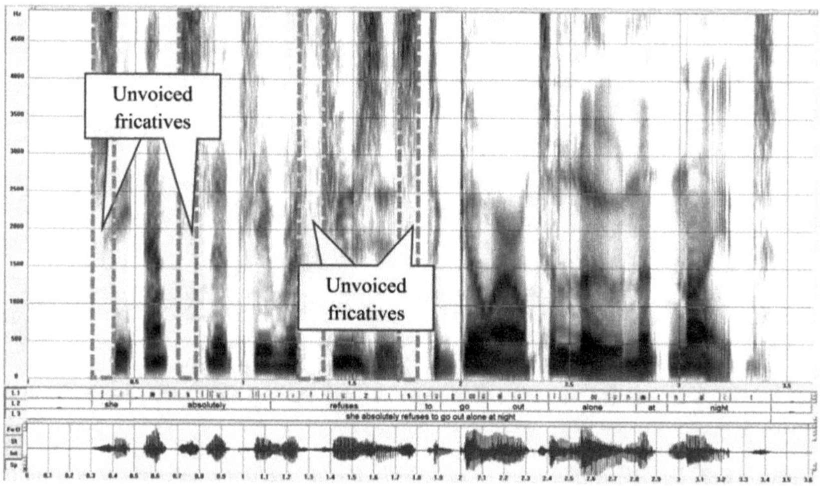

Figure 6.7. *Locating unvoiced fricatives [ʃi:æbsllu:tlɪrɪfju:zɪstʊgo ʊʊal ʊtllo ʊʊnætnali:t]*

Only voiced fricatives have harmonics for the low frequencies of the spectrum, around 120 Hz for male voices and 200 Hz to 250 Hz for female voices (Figures 6.8 and 6.9).

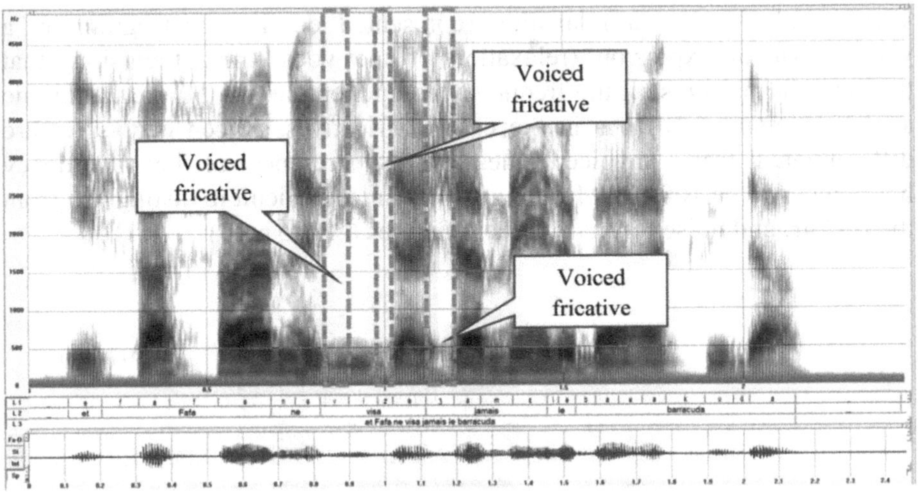

Figure 6.8. *Locating voiced fricatives [efafanəvizaʒamɛləbarakyda]*

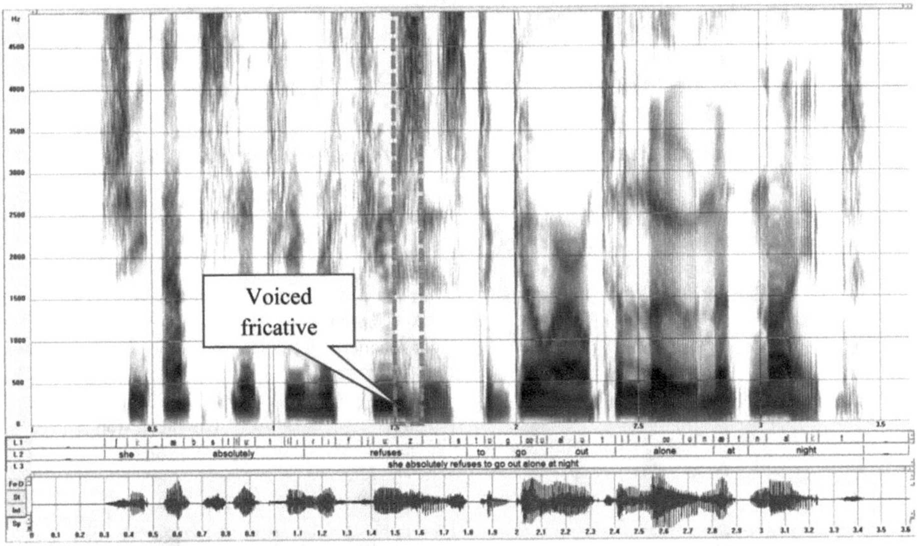

Figure 6.9. *Locating voiced fricatives [ʃi:æbsllu:tlırıfju:zıstʊgoʊʊalʊtlloʊʊnætnali:t]*

6.2.7. *Occlusives, stop consonants*

The next step is the occlusives. The voiced and unvoiced occlusives are characterized by the holding, the closure of the vocal tract appearing as a silence, followed by a relaxation represented on the spectrogram by a vertical bar of explosion (relaxation of the vocal tract occlusion) that theoretically presents all the frequency components in the Fourier harmonic analysis, Figure 6.10). As in the case of fricatives, voiced occlusives are differentiated from unvoiced ones by the presence of low frequency harmonics, in the 100 –200 Hz range, sometimes difficult to distinguish from the background noise for male voices (Figures 6.10 to 6.13).

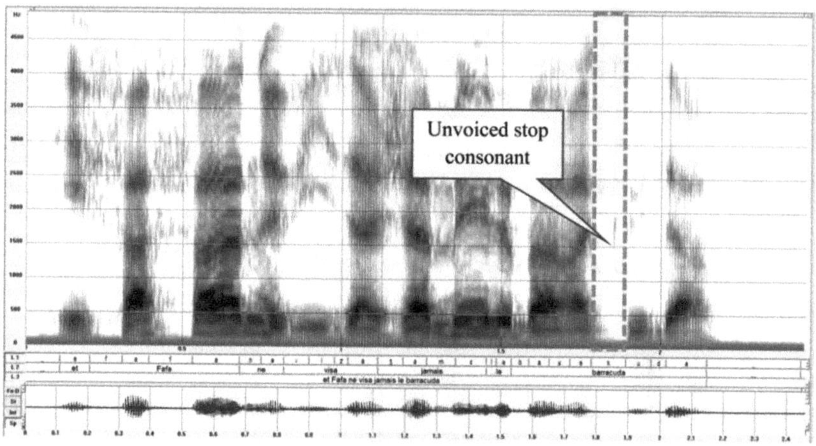

Figure 6.10. *Locating unvoiced stop consonants [efafanəvizaʒamɛləbarakyda]*

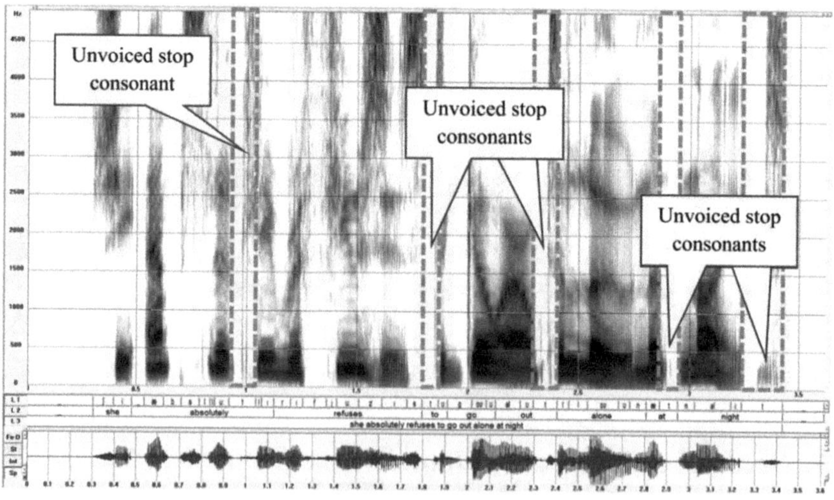

Figure 6.11. *Locating unvoiced stop consonants [ʃi:æbsllu:tlırıfju:zıstʊgo ʊʊal ʊtllo ʊʊnætnali:t]*

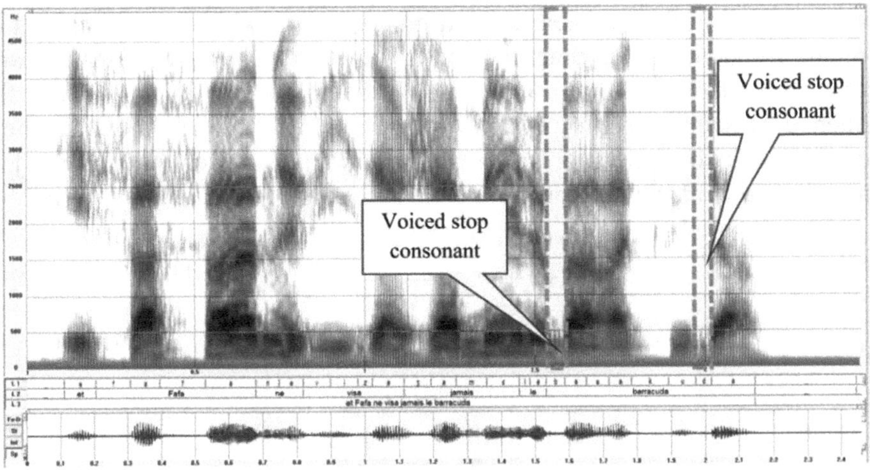

Figure 6.12. *Locating voiced stop consonants [efafanəvizaʒamɛləbarakyda]*

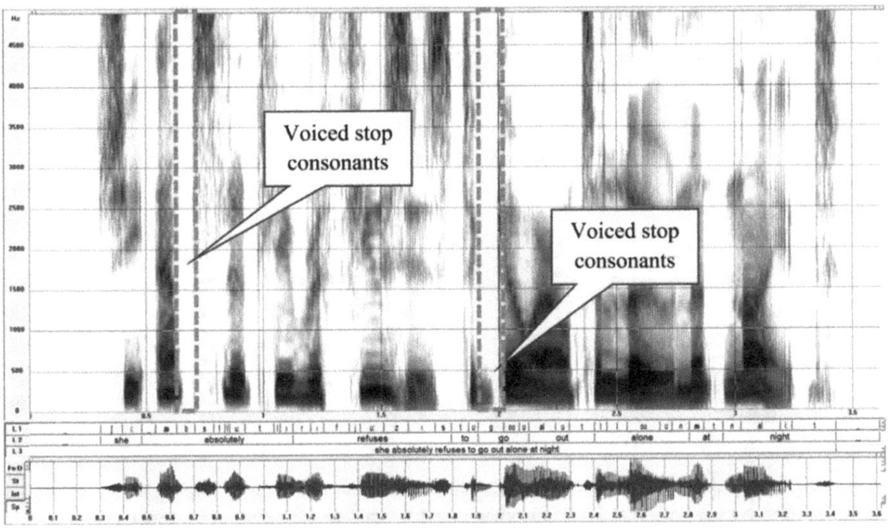

Figure 6.13. *Locating voiced stop consonants [ʃi:æbsllu:tlırıfju:zıstʊgoʊʊalʊtlloʊʊnætnali:t]*

6.2.8. *Vowels*

We then move on to vowel segmentation. Vowels have a specific harmonic structure, but it is difficult to differentiate between them at the outset without, at the very least, resorting to an approximate measurement of formant frequencies. In practice, the prior segmentation of fricative and occlusive consonants often makes this identification useless, since the sequence of sounds in the phonetic transcription is known, provided there are no vowels contiguous or in contact with a nasal consonant [m] or [n] or with a lateral consonant [l] or [r].

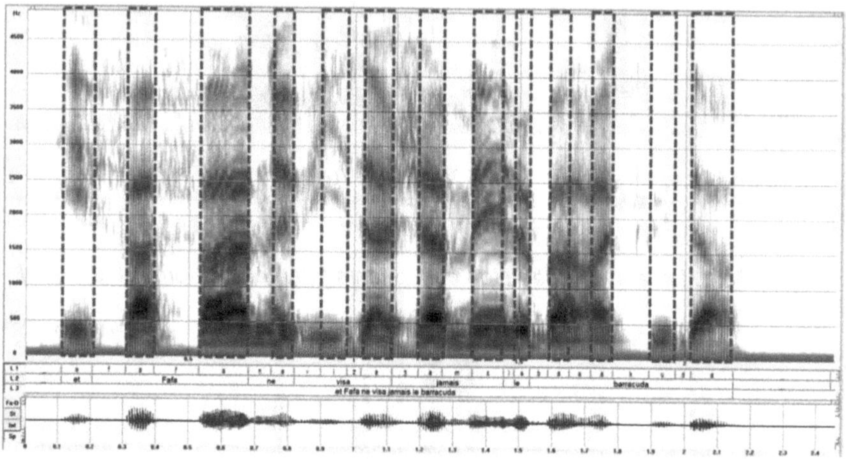

Figure 6.14. *Locating vowels [efafanəvizaʒamɛləbarakyda]*

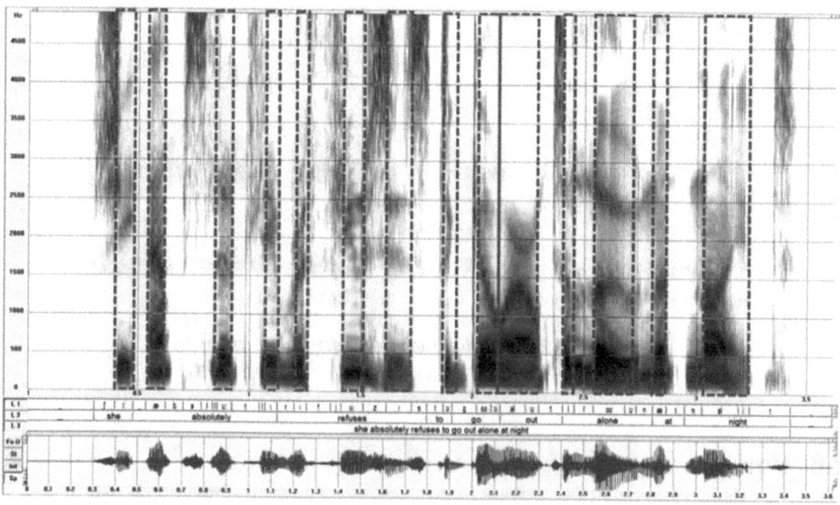

Figure 6.15. *Locating vowels and diphthongs [ʃi:æbsllu:tlırıfju:zıstʊgoʊʊalʊtlloʊʊnætnali:t]*

In the latter case, one must rely on the relative stability of the formants that the vowels are supposed to present (in French). Figures 6.14 and 6.15 show an example of a vowel sequence. In general, vowels are also characterized by a greater amplitude than consonants on oscillograms, as can be seen (in blue in the figures).

6.2.9. *Nasals*

Nasal consonants are often the most difficult to segment. They generally have a lower amplitude than the adjacent vowels, resulting in lower intensity formants. However, they can often be identified by default by first identifying the adjacent vowels.

The same is true for liquids [l] and variants of [r], [R] (which can also be recognized by the visible wideband flaps with sufficient time zoom, Figures 6.16 and 6.17).

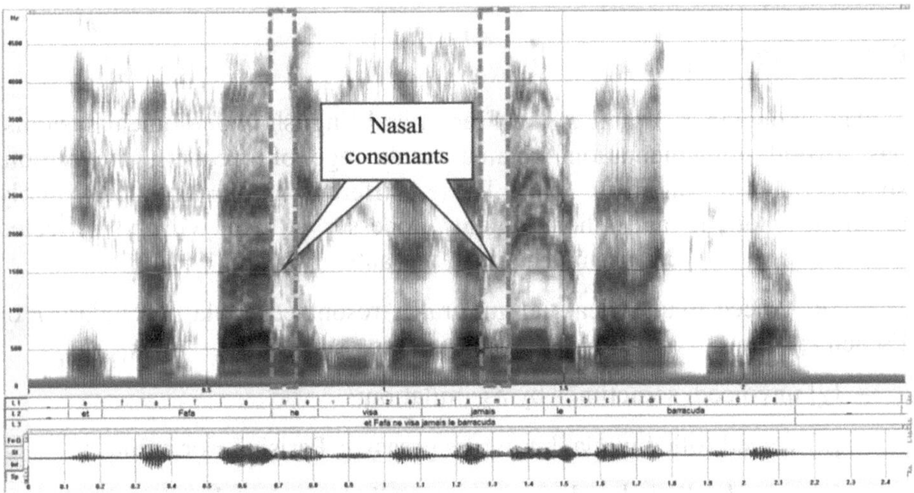

Figure 6.16. *Locating nasal consonants [efafanəvizaʒamɛləbarakyda]*

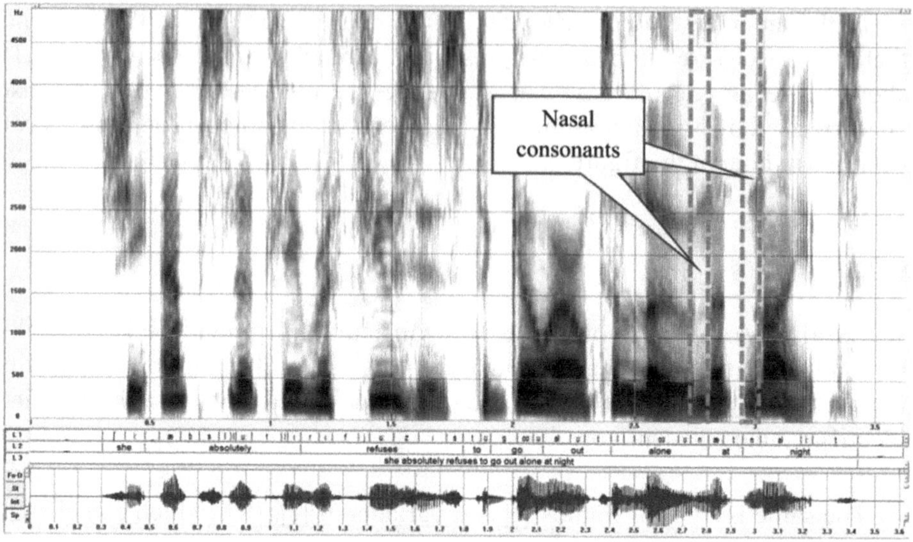

Figure 6.17. *Locating nasal consonants [ʃi:æbsllu:tlırıfju:zıstʊgoʊʊalʊtlloʊʊnætnali:t]*

6.2.10. *The R*

The phoneme |r|, a unit of the French and English phonological systems, has several implementations related to the age of speakers, the region, socio-economic variables, etc. These different implementations imply different phonetic and phonological processes, which will be reflected differently on a spectrogram. The main variants of |R| are:

– [ʁ] voiced uvula fricative, guttural *r*, resonated uvula *r*, standard *r* or French *r*;

– [χ] unvoiced, sometimes by assimilation;

– [ʀ] voiced rolled uvula, known as greasy *r*, Parisian *r* or uvula *r*;

– [r] voiced rolled alveolar, known as rolled *r* or dental *r*;

– [ɾ] voiced beaten alveolar, also known as beaten *r*;

– [ɻ] voiced spiral retroflex, known as retroflex *r* (especially in English and in some French speaking parts of Canada).

6.2.11. *What is the purpose of segmentation?*

Syllabic and telephone segmentation is a recurrent activity in speech research and development, which, until recently, was carried out manually on relatively short recordings. However, manual segmentation carried out by visual inspection of spectrograms by trained experts is a very long process, not only requiring expertise in acoustics but also in the phonetic and phonological specificities of the language under consideration. The sheer cost of manual segmentation is the main incentive for developing automatic speech segmentation, as the analysis of very large corpora of spontaneous speech is becoming a key research topic in the field of linguistics, with important applications in speech recognition and synthesis.

6.2.12. *Assessment of segmentation*

The effectiveness of an automatic segmentation method is usually assessed by comparison with a manual segmentation of the same records. Despite often published claims, their reliability only seems acceptable for recordings of fairly good quality (in other words, with a high signal-to-noise ratio, no echo, low signal compression, wide frequency range, etc.). In addition, most of these methods are language-specific and require separate adaptation of phone models for each language considered, preventing their use in foreign language teaching applications, where learner performance may differ considerably from the norm.

6.2.13. *Automatic computer segmentation*

Automatic speech segmentation errors are often listed as the number of insertions and deletions from a reference that are usually obtained by expert visual inspection of spectrograms. It is customary to add statistics to these values, such as the mean and standard deviation of the time differences of the boundaries for the corresponding segments, as well as some indication of the distribution of these differences. When using forced alignment, algorithms such as EasyAlign[1] and WebMaus[2] rely on similar IPA spelling

1 http://latlcui.unige.ch/phonetique/easyalign.php.

2 https://clarin.phonetik.uni-muenchen.de/BASWebServices/interface.

conversion systems and will therefore have no, or very few, insertions or deletions, and these occasional differences only reflect the quality of the IPA conversion, as assessed by a phonological expert and phonetic knowledge (see below). It is therefore expected that non-standard pronunciation, deviating from the standard implied in the system's text-IPA conversion, will create more errors of this type, although the underlying phonetic models can be adapted to specific linguistic varieties.

Despite these discouraging considerations, given the scope of the task of segmenting oral corpora of ever-increasing duration, and given the increasingly considerable demand from new users such as Google, Microsoft, IBM, Amazon, etc., to train deep learning speech recognition and synthesis systems, several automatic segmentation algorithms have recently been developed. Even if these systems are imperfect, they considerably reduce the work of segmentation specialists on spectrograms, even if it means manual verification and correction. Among the most widely used processes, if not the most well-known, are EasyAlign, WebMaus and Astali, all based on a probabilistic estimation of the spectral characteristics of phones, as determined from the spelling transcription into IPA notation (or equivalent) (Figure 6.11). A different process based on the comparison of a synthesized sentence with the corresponding original speech segment is implemented in the WinPitch software.

Current methods of automatic speech segmentation make use of many signal properties: analysis from data provided by intensity curves (convex shell, hidden Markov models, HMM Gaussian modeling), a combination of periodicity and intensity, multiband intensity, spectral density variations, forced alignment of phonetic transcription, neural network, hybrid centroid, among many others (Figure 6.18).

For EasyAlign, for example (Goldman 2020), the IPA transcription is performed from a linguistic analysis of the spelling transcription (identification of sequences and links) in order to produce a phonetic transcription from a dictionary and the rules of pronunciation. The system uses a pronunciation dictionary for a speech recognition algorithm operating by forced alignment. Additional rules are used to identify syllable boundaries based on the principle of sonority variation.

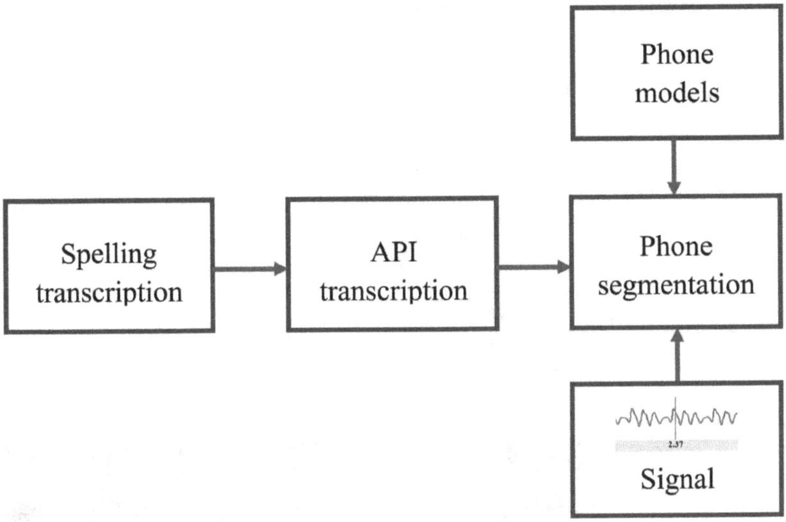

Figure 6.18. *Schematic diagram of automatic segmentation into phones*

6.2.14. *On-the-fly segmentation*

The principle of on-the-fly segmentation consists of an operator clicking on spelled units, syllables, words, groups of words or phrases with a mouse as they are identified when played back. Since such an operation is difficult to perform in real time, especially for the smallest units, a programmable slowed-down playback is available to ensure that hearing coordination and proper positioning of the graphical cursor are coordinated under good conditions. Most of the problems inherent to automatic segmentation, such as background noise, dialectal variations, echo, etc., are then entrusted to a human operator, *a priori* more efficient than a dedicated algorithm. A set of ergonomic functions, including back boundary erasing, retakes, mouse speed variation, etc., make the whole system very efficient, especially since, unlike automatic systems, the operator remains in control of the process and can detect possible spelling transcription errors at the same time, which is rarely done in practice for very large corpora (Figure 6.19).

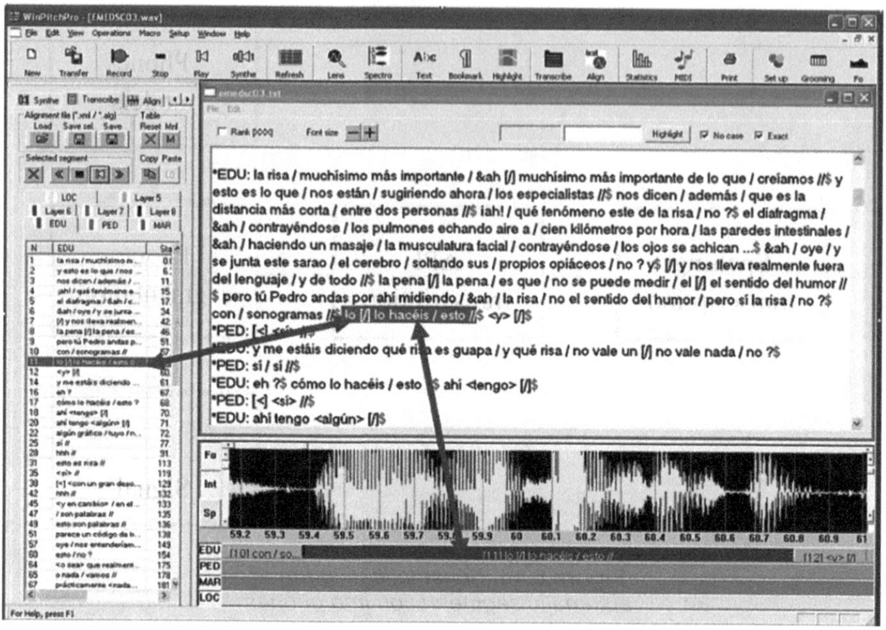

Figure 6.19. *Example of on-the-fly segmentation with slowdown of the speech signal*

6.2.15. *Segmentation by alignment with synthetic speech*

The principle of segmentation by alignment with synthetic speech is based on the forced alignment of the recording to be segmented with the text-to-speech synthesis of the annotated text (Malfrère and Dutoit 1997). By retrieving the temporal boundaries of the phones produced by the synthesizer, which are accessible among the operating system temporal semaphores, the corresponding boundaries in the signal to be segmented are found by forced alignment. The orthographic-phonetic transcription, the specific features of the languages considered, are thus (normally) taken into account by the text-to-speech synthesis available in many operating systems (Windows, MacOS, etc.).

The segmentation is done in two phases: generation of the spelled text followed by a forced Viterbi dynamic comparison alignment, according to the following steps (Figure 6.20):

1) dividing the time axes into segments (for example, 50 ms or 100 ms) forming a grid (Figure 6.20);

2) selecting a comparison function between wideband spectra (for example, the sum of amplitudes at logarithmically spaced frequencies);

3) starting from the 0.0 box (bottom left of the grid);

4) comparing the similarity obtained by moving to the 1.0, 1.1 and 0.1 boxes, and moving to the box giving the best similarity between the corresponding spectra, according to the function chosen in (2);

5) this is done until the upper right corner (box n, m) is reached;

6) finally, retracing the path travelled by the Tom Thumb strategy from the memorization of each box movement on the grid.

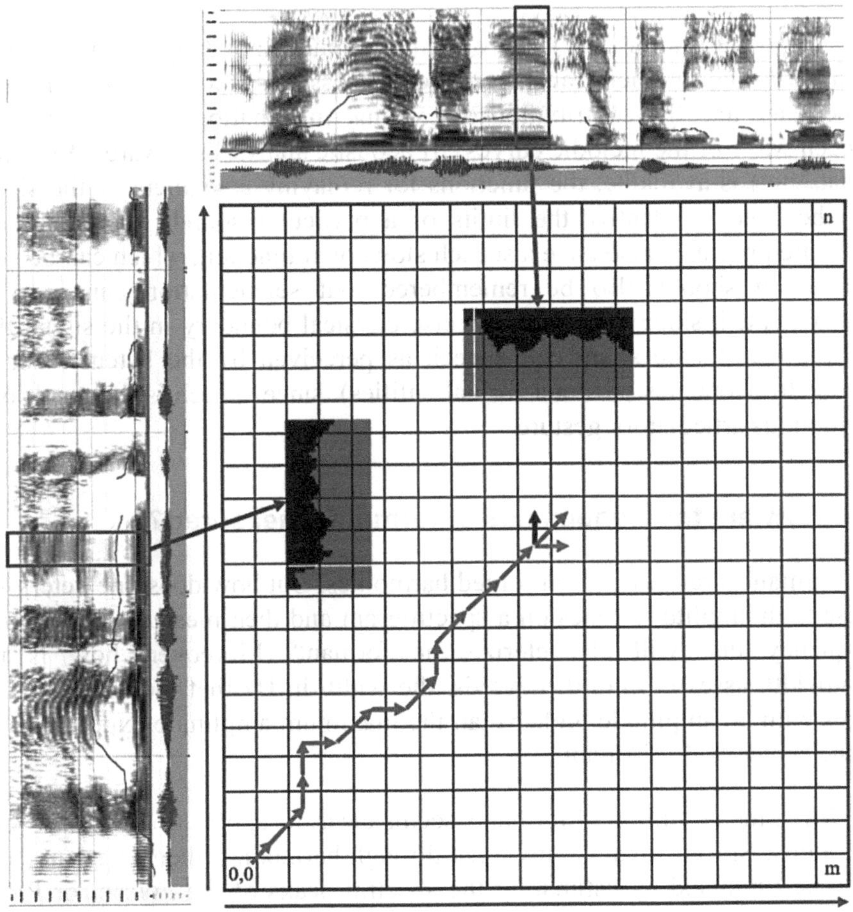

Figure 6.20. *Alignment by dynamic comparison (Dynamic Time Warping, DTW)*

Efficiency depends on the comparison function between spectra and the step chosen for each axis. A fine grid makes it possible to better consider the variations in speech rate between the two spectra, but leads to longer calculations. On the other hand, the temporal resolution of the spectra is particularly critical, so as to base the comparison on formant zones rather than on harmonics, which are much more variable than formants if the two texts at least correspond in terms of spelling transcription. Alignment between a male and a female voice will often require different settings for each spectrum.

6.2.16. *Spectrogram reading using phonetic analysis software*

The segmentation of a spectrogram combines knowledge of articulatory characteristics and their translation over successive spectra. It also calls upon experience and the contribution of external information (for example, the speech wave or the melodic curves). If acoustic analysis software (WinPitch, Praat, etc.) is available, the functions for replaying a segment of the signal can be used to identify the limits of a particular sound. Some of these software programs also have a speech slowdown function, which can be very helpful. It should also be remembered that segmentation can only be approximated, since there is no precise physical boundary in the signal that corresponds to the sounds of speech as perceived by the listener, nor of course to phonemes (abstract formal entities), since speech is the result of a continuous articulatory gesture.

6.3. How are the frequencies of formants measured?

Formants are areas of reinforced harmonics. But how does one determine these high amplitude zones on a spectrogram and then measure their central frequency that will characterize the formant? Moreover, how is the bandwidth assessed, in other words, the width in Hz that the formant zone takes with an amplitude higher than the maximum amplitude expected to be at the center of the formant?

While the definition of formants seems clear, its application raises several questions. Since formants only exist through harmonics, their frequency can only be estimated by estimating the spectral peaks of the Fourier or Prony spectra (Figure 6.21).

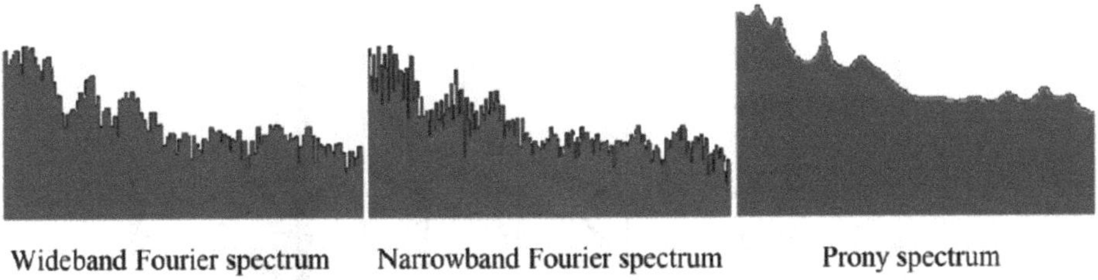

Figure 6.21. *Vowel [a]: Fourier spectrum wideband, narrowband and Prony's spectrum*

Fourier analysis requires, on the one hand, a visual estimation of spectral peaks, which is not always obvious, and on the other, introduces an error (at least by visual inspection) that is equal to half the difference between two harmonics close to the spectral maximum. It is, however, possible to reduce this error by parabolic interpolation. When the Fourier spectrum is narrowband, so as to better distinguish the harmonics – which is, let us not forget, at the cost of a lower temporal resolution and therefore at the cost of obtaining an average over the entire duration of the time window necessary for the analysis – visual estimation is not necessarily easier. Moreover, the development of computer-implemented algorithms to automate this measurement proves very difficult, and the existing realizations are not always convincing.

An extreme case of the formant measure is that put forward by the "singer's problem". Imagine that a soprano must sing the vowel [ə] in her score, whose formants are respectively F1 500 Hz, F2 1,500 Hz, F3 2,500 Hz, F4 3,500 Hz, etc. As soon as the singer's laryngeal frequency exceeds the frequency of the first formant, such as when reaching 700 Hz, this first formant can no longer be performed since no harmonic will correspond to 500 Hz, the frequency of the first formant of the vowel to be performed (Figure 6.16). This illustrates the separation between the laryngeal source, responsible for the frequency of the harmonics, and the vocal tract configuration, responsible for the frequency of the formants (in reality, an interaction between the two processes exists but can be neglected as a first approximation).

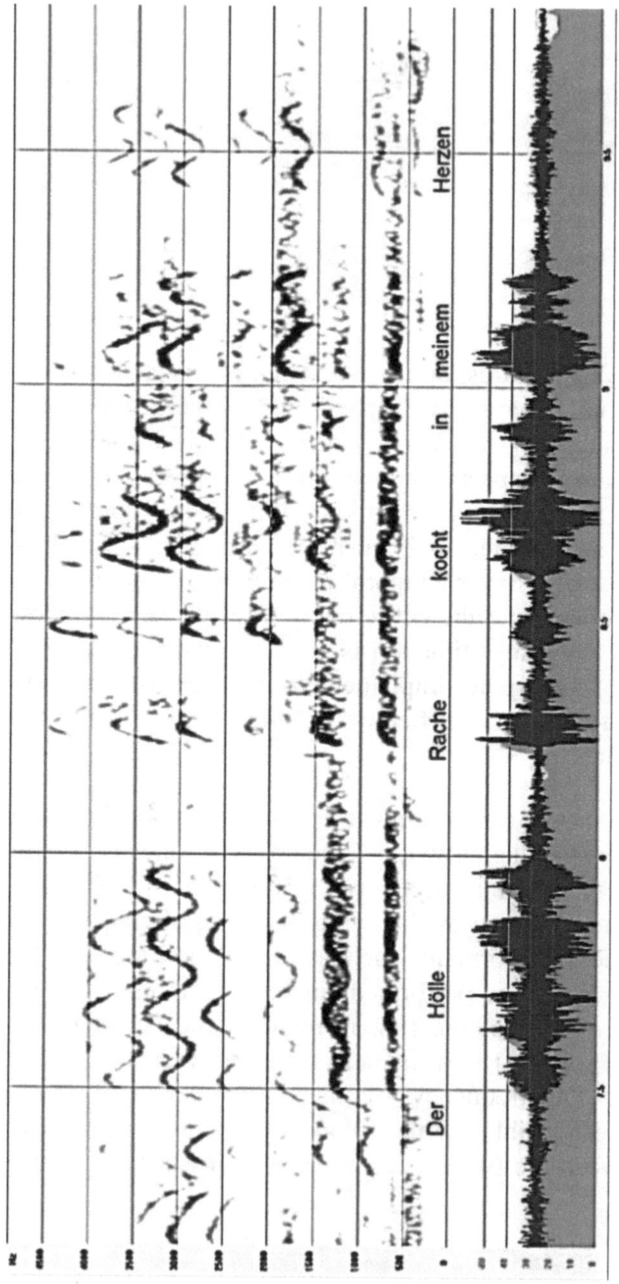

Figure 6.22. *Spectrogram from the beginning of the "Air de la Reine de la nuit" by the singer Natalie Dessay:* Der Hölle Rache kocht in meinem Herzen *(Hell's vengeance boils in my heart)*

Prony's analysis seems much more satisfactory for measuring formants, in that it presents peaks corresponding to formants that are easy to identify, both visually and by an algorithm. However, the position of these peaks depends on the number of coefficients in the calculation of the linear prediction coefficients. Figures 6.23 and 6.24 illustrate the effect of the filter order on the resulting spectrum. Tables 6.2 and 6.3 give the formant frequencies corresponding to the peaks in each spectrum.

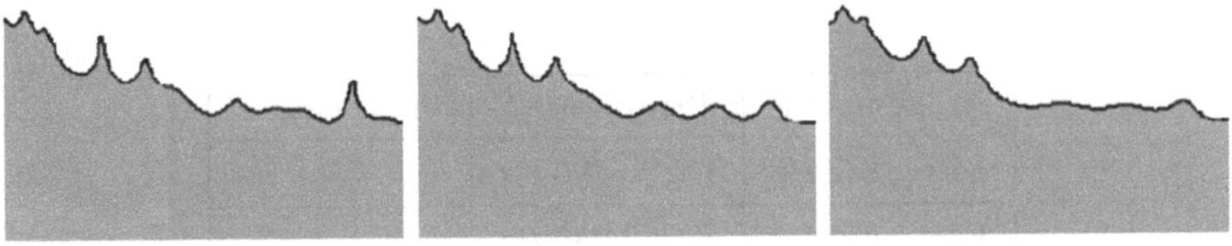

Prony order 12 Prony Order 10 Prony order 8

Figure 6.23. *Prony spectrum of order 12, 10 and 8*

Order 12		Order 10		Order 8	
F1	689 Hz	F1	689 Hz	F1	545 Hz
F2	1,263 Hz	F2	1,263 Hz	F2	1,033 Hz
F3	2,641 Hz	F3	2,641 Hz	F3	2,670 Hz
F4	3,875 Hz	F4	3,875 Hz	F4	3,990 Hz
F5	4,656 Hz	F5	4,737 Hz	F5	6,373 Hz
F6	6,747 Hz	F6	6,632 Hz	F6	8,125 Hz
F7	8,240 Hz	F7	8,240 Hz	F7	9,790 Hz
F8	9,532 Hz	F8	9,646 Hz	–	
F9	10,450 Hz	F9	10,370 Hz	–	
F10	undetectable	–		–	
F11	undetectable	–		–	

Table 6.2. *Peak values of the Prony spectrum of order 12, 10 and 8*

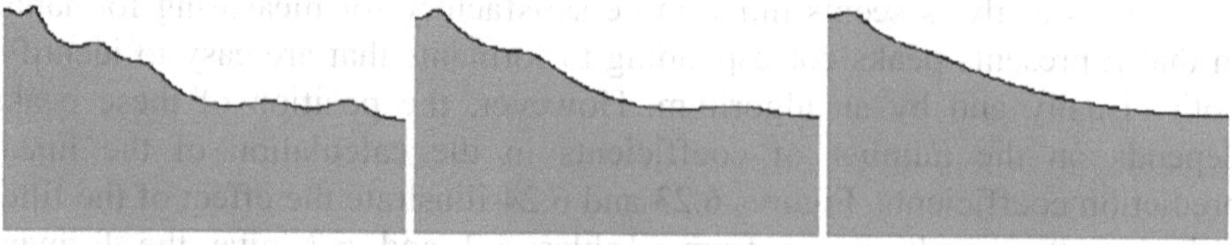

Figure 6.24. *Prony spectrum of order 6, 4 and 2*

Prony order 6	Prony order 4	Prony order 2
F1 832 Hz	F1 796 Hz	F1 568 Hz
F2 2,670 Hz	F2 3,730 Hz	–
F3 3,933 Hz	F3 9,388 Hz	–
F4 7,350 Hz	–	–
F5 9,503 Hz	–	–

Table 6.3. *Peak values of the Prony spectrum of order 6, 4 and 2*

These examples show the influence of the filter order, an analysis parameter generally accessible in the analysis software, on the formant values obtained.

The values of the peaks that are supposed to represent the formants seem to have stabilized after a sufficient order of calculation. What does this reflect? To answer this question, let us increase the Prony order considerably, for example up to 100, and we will then see peaks appear that no longer correspond to the formants, but to the harmonic frequencies. The local vertices of these peaks are similar to those of the Fourier analysis and do indeed correspond to the formant frequencies of the analyzed vowel (Figure 6.25).

In summary, assuming a (relatively) stationary signal, the number of formants being equal to the number of coefficients/2, we can adopt the heuristic rule:

Number of LPC coefficients = 2 + (Sampling frequency/1,000)

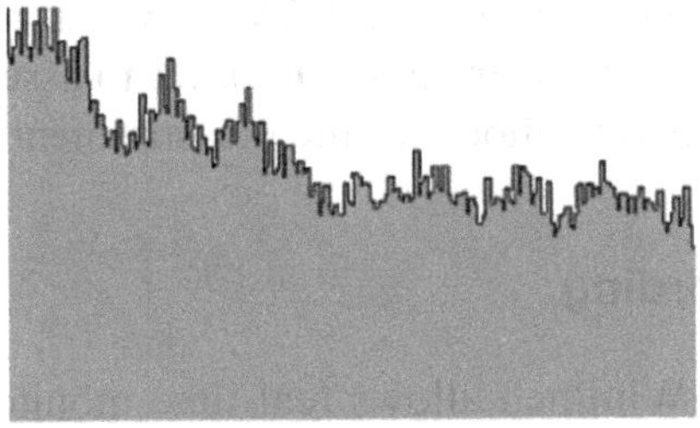

Figure 6.25. *Prony spectrum of order 100 showing peaks corresponding to the harmonics of the signal*

It should also be noted that the precise position of the formants depends on the method of resolution (autocorrelation, covariance, Burg, etc.), and also that the method is not valid for occlusives or nasals (unless the ARMA model is used, the resolution of which is generally not available in phonetic analysis software).

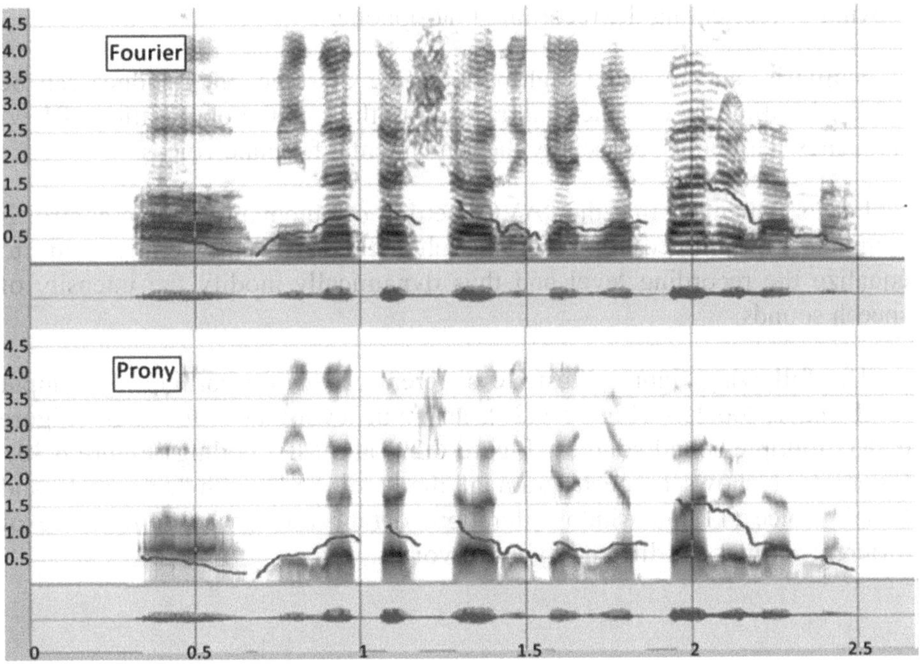

Figure 6.26. *Comparison of Fourier and Prony spectrograms*

To conclude this chapter, Figure 6.26 shows a comparison of Fourier (mediumband) and Prony spectrograms in order to assess the advantages and disadvantages of the two methods for the measurement of formants.

6.4. Settings: recording

Software such as WinPitch allows real-time monitoring of the recording by not only displaying the speech wave curve (a waveform representing sound pressure variations as perceived by the microphone), but also a wideband or narrowband spectrogram, as well as the corresponding melodic curve. This information allows the operator to correct the recording parameters, particularly the microphone position if necessary, by reporting back to the speaker. The visual identification of noise sources is facilitated by their characteristic print on the narrowband spectrographs. This is because, unlike the recorded speech source, the harmonics of the noise sources generally have a constant harmonic frequency ratio. It is therefore easy to recognize them and to eliminate or reduce them, either by neutralizing the source (for example, by switching off a noisy engine) or by moving or reorienting the recording microphone.

Similarly, an unwanted high-pass setting of some microphones can be detected in time, before the final recording, through the observation of low-frequency harmonics, especially the fundamental for male voices.

Let us remember that the "AVC" setting of automatic volume control is completely prohibited for recording phonetic data. This setting tends to equalize the recording level and thus dynamically modify the intensity of speech sounds.

The following figures illustrate different cases of recording at an input level that is too low (Figure 6.27), an input level that is too high (Figure 6.28), multiple sound sources (Figure 6.29) and MP3 coding (Figure 6.30). Figure 6.31 shows a spectrogram of the same bandwidth for a male and female voice (11 ms window duration) demonstrating the dependence of different settings for different types of voices.

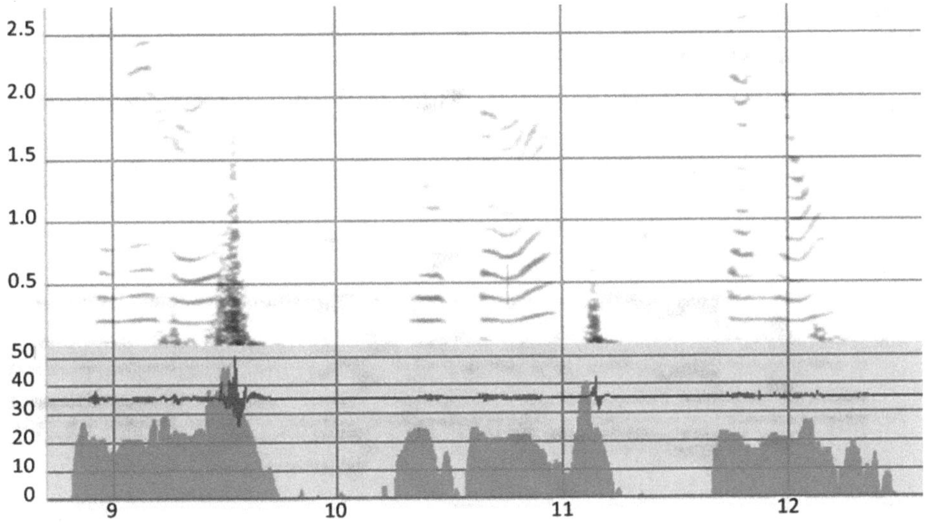

Figure 6.27. *Example of a recording level that is too low: the harmonics of the recorded voice are barely visible*

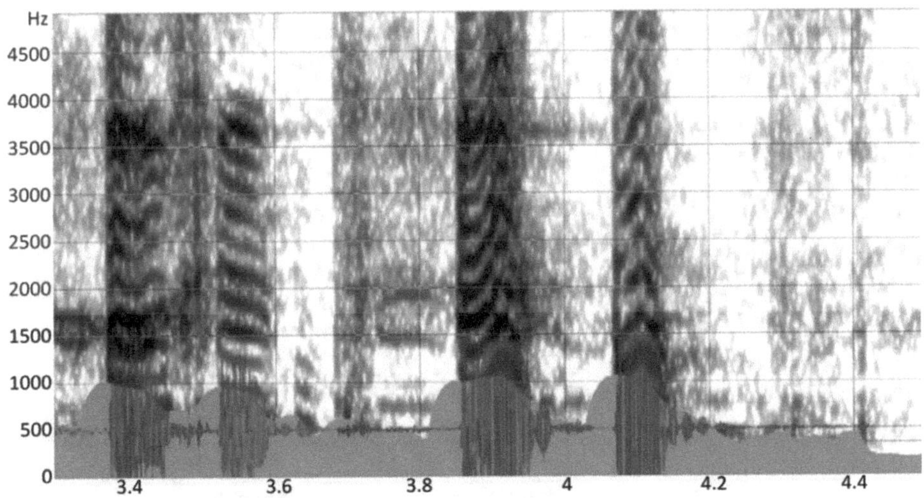

Figure 6.28. *Example of a recording level that is too high: harmonic saturation is observed on the narrowband spectrogram*

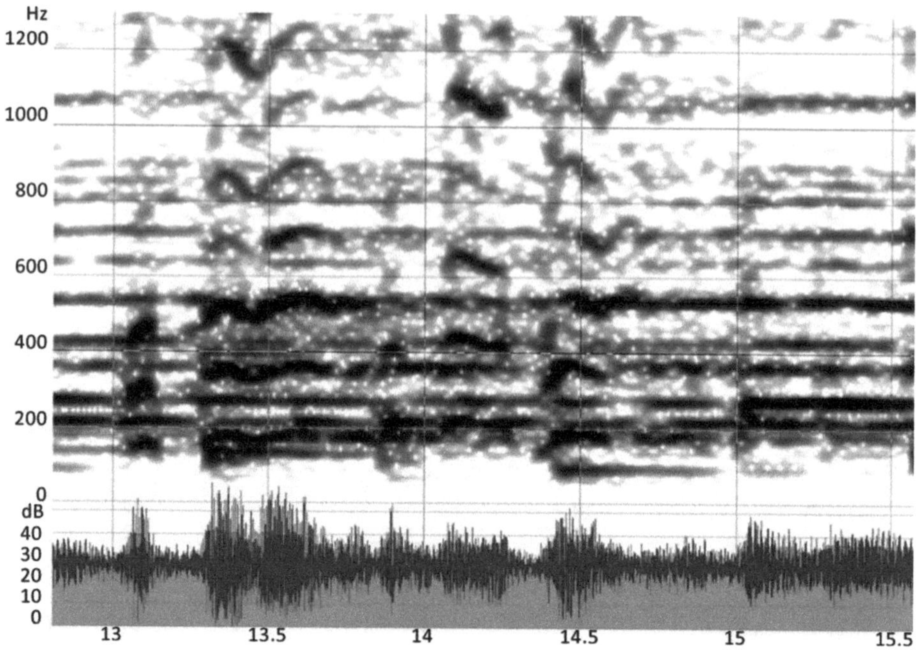

Figure 6.29. *Presence of constant frequency noise harmonics (musical accompaniment) that is superimposed on the harmonics of the recorded voice*

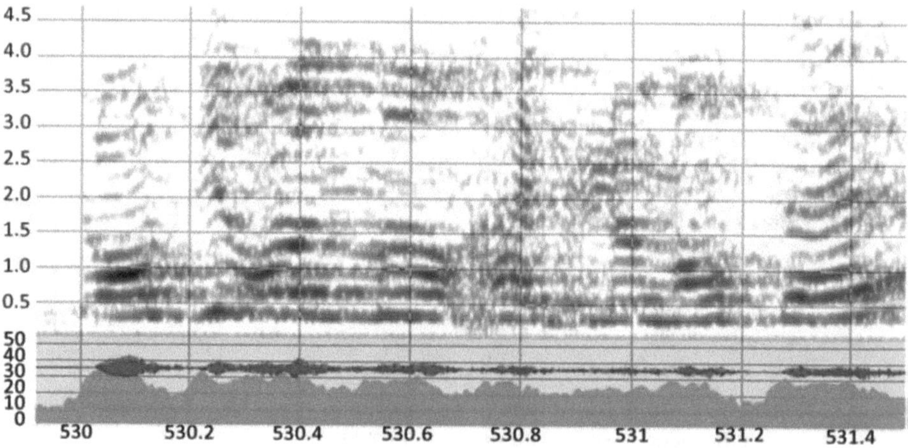

Figure 6.30. *Effect of MP3 encoding-decoding on the narrowband representation of harmonics*

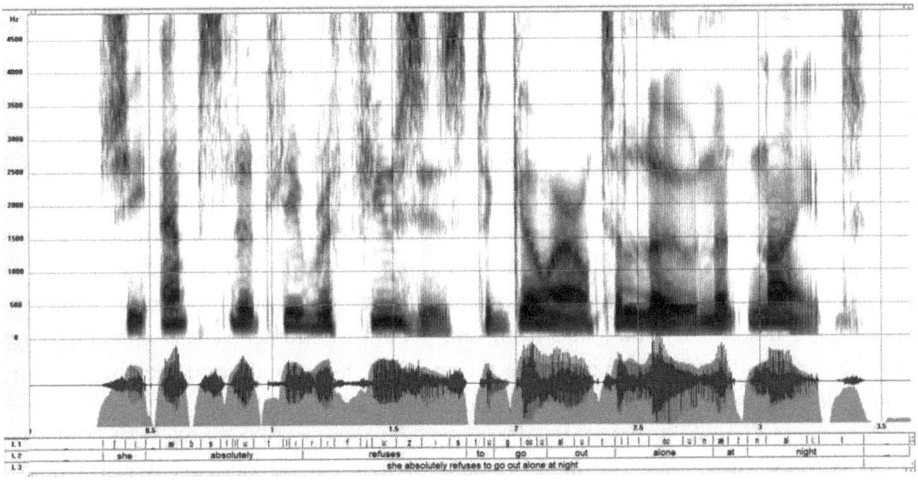

a)

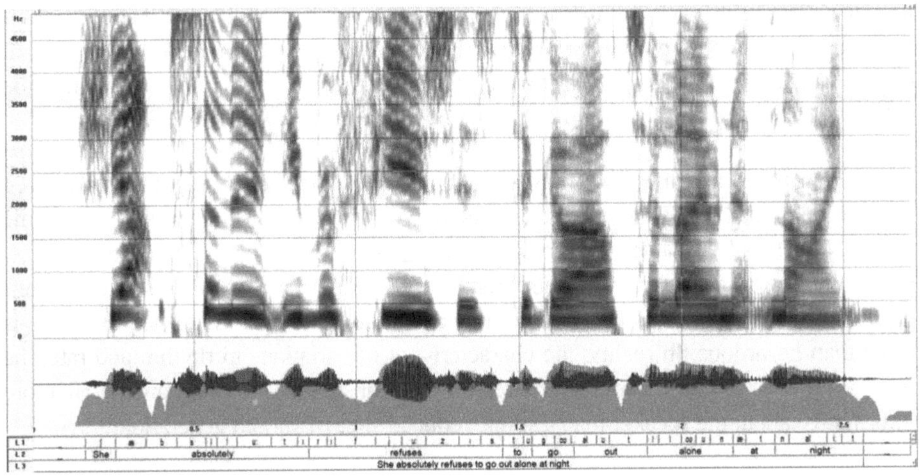

b)

Figure 6.31. *Spectrograms at the same bandwidth for a male and female voice (11 ms window duration) of the sentence "She absolutely refuses to go out alone at night" (Corpus Anglish)*

When the first spectrographs for acoustic speech analysis appeared, law enforcement agencies in various countries (mostly in the US and USSR at the time) became interested in their possible applications in the field of suspect identification; so much so that US companies producing these types of equipment took on names such as Voice Identification Inc. to better ensure their visibility to these new customers.

Most of the time, the procedures put in place were based on the analysis of a single vowel, or a single syllable, which obviously made the reliability of these voice analyses quite random. Indeed, if we can identify the voices of about 50 or, at most, 100 relatives, it is difficult for us to do so before having heard a certain number of syllables or even whole sentences. Moreover, unlike fingerprints or DNA spectra, the spectrum and formant structure of a given speaker is related to a large number of factors such as physical condition, the rate of vocal cord fatigue, the level of humidity, etc. On the other hand, voiceprint identification implies that certain physical characteristics of the vocal organs, which influence the sound quality of speech, are not exactly the same from one person to another. These characteristics are the size of the vocal cavities, the throat, nose and mouth, and the shape of the tongue joint muscles, the jaw, the lips and the roof of the mouth.

It is also known that the measurement of formant frequencies is not a trivial operation. In the case of Fourier analysis, it depends on the expertise of the observer to select the right bandwidth, locate the formants appropriately, and estimate their frequency, regardless of the laryngeal frequency. This last condition gives rise to many errors in the calculation of the formants, since their frequency must be estimated from the identification of a zone of higher amplitude harmonics.

In fact, the similarities that may exist for spectrograms corresponding to a given word, pronounced by two speakers, may be due to the fact that it is precisely the same word. Conversely, the same word spoken in different sentences by the same speaker may have different patterns. On the spectrogram, the repetitions show very clear differences. The identification of a voice over a restricted duration, a single vowel for example, is more than hazardous. In reality, the characteristics of variations in rhythm and rate, the realization of melodic contours on stressed (and unstressed) vowels are much more informative about the speaker than a single reduced slice of speech generation.

This has led to a large number of legal controversies, which have had a greater impact in Europe than in the United States. In the United States (as in other countries), it is sometimes difficult for scientists specializing in speech sound analysis to resist lawyers who are prepared to compensate the expert's work very generously, provided that their conclusions point to the desired direction. As early as 1970, and following controversial forensic evaluations, a report by the Acoustical Society of America (JASA) concluded that identification based on this type of representation resulted in a significant and hardly predictable error rate (Boë 2000).

Box 6.1. *The myth of the voiceprint*

7

Fundamental Frequency and Intensity

7.1. Laryngeal cycle repetition

The link between voice pitch and vocal fold vibrations was among the first observed on the graphic representation of speech sound vibrations as a function of time. Indeed, the near repetition of a sometimes-complex pattern is easy to detect. For example, the repetitions of a graphical pattern can easily be recognized in Figure 7.1, translating the characteristic oscillations of laryngeal vibrations for a vowel [a].

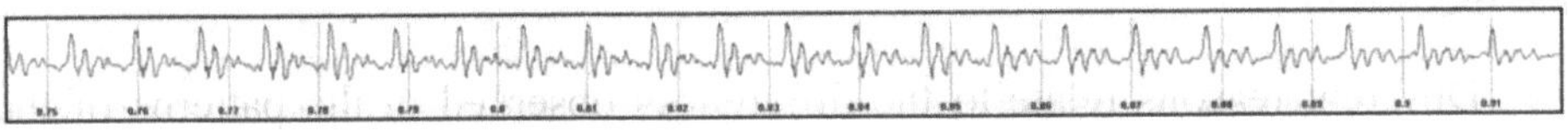

Figure 7.1. *Repeated characteristic laryngeal vibration pattern as a function of time for a vowel [a]*

Looking at Figure 7.1 in more detail (vowel [a], with the horizontal scale in seconds), there are 21 repetitions of the same pattern in approximately (0.910 – 0.752) = 0.158 seconds. It can be deduced that the average duration of a vibration cycle is about 0.158/21 = 0.00752 seconds, which corresponds to a laryngeal frequency of 1/0.00752 = about 132 Hz.

In another occurrence of the vowel [a], pronounced by the same speaker in the same sentence, a comparable pattern can be seen. However, the vibrations following the main vibration are more significant in the first example than in the second, and if we take a particularly close look, we observe that some

For a color version of all the figures in this chapter, see: www.iste.co.uk/martin/speech.zip.

peaks present several close bounces of equal amplitude around the 1.74–1.77 interval.

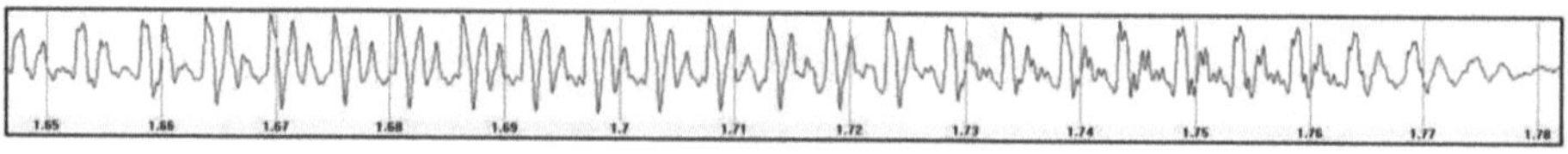

Figure 7.2. *Patterns for another occurrence of vowel [a]*

For the vowel [i] in Figure 7.3, there are 23 repetitions of a somewhat different oscillation pattern of about (1,695 − 1,570)/23 = 5.42 ms, mean period, or about 184 Hz.

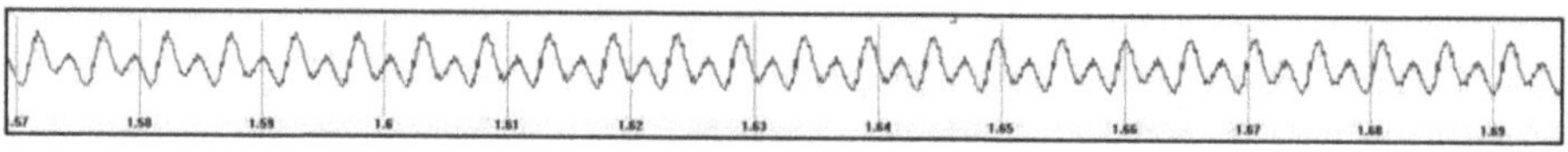

Figure 7.3. *Patterns for an occurrence of vowel [i]*

The patterns repeated in each period are not exactly reproduced identically, due to small, unavoidable variations in the configuration of the vocal tract during phonation, leading to changes in the phase of the harmonic components.

Different reasons relate to the differences observed in the patterns of the two examples in [a] and [i]. In the early days of research on speech acoustics, it was thought that the description of the pattern could be sufficient enough to characterize vowels such as [a] and [i]. Unfortunately, this is not the case!

If, as we saw in the chapter devoted to the spectral analysis of vowels, the harmonic components created by laryngeal vibrations have relatively large amplitudes in certain frequency ranges (the formants) for each vowel – areas resulting from the articulatory configuration to produce these vowels – the relative phases of the different harmonics are not necessarily stable and can not only vary from speaker to speaker, but also during the emission of the same vowel by a single speaker. The addition of the same harmonic components, but shifted differently in phase, may give different waveforms, while the perception of the vowel does not change since it is resistant to phase differences.

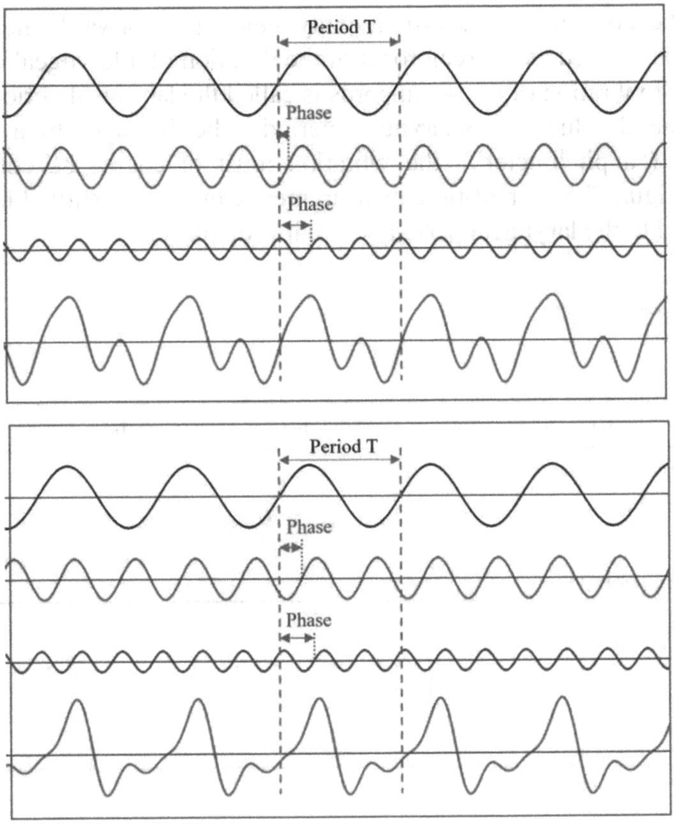

Figure 7.4. *Effect of phase change on three harmonic components added with different phases (top and bottom of figure)*

7.2. The fundamental frequency: a quasi-frequency

Theoretically, the speech fundamental frequency is not a frequency, and the periods observable on the signal (on the waveform, also called the oscillographic curve) are not periods, given that the repeated patterns are not identical in every cycle. However, the definition and concept of frequency necessarily refers to strictly periodic events. We then speak about quasi-periodic events, and this idealization – or compromise with reality – proves more or less acceptable in practice. It may lead to misinterpretations of the results of acoustic analysis that are based on the hypothesis of periodicity, a hypothesis which, in any case, is never fully verified.

With the concept of quasi-periodicity being put forward, the laryngeal frequency is defined as the reciprocal of the duration of a laryngeal cycle, thus one cycle of vibration of the vocal cords is called the laryngeal period. In order for a measured value of this laryngeal period to be displayed by a measuring instrument ("a pitchmeter"), the vibration must of course be complete, as shown in Figure 7.5. A real-time measurement can only be shifted by a period that is equal to the laryngeal period before it is displayed.

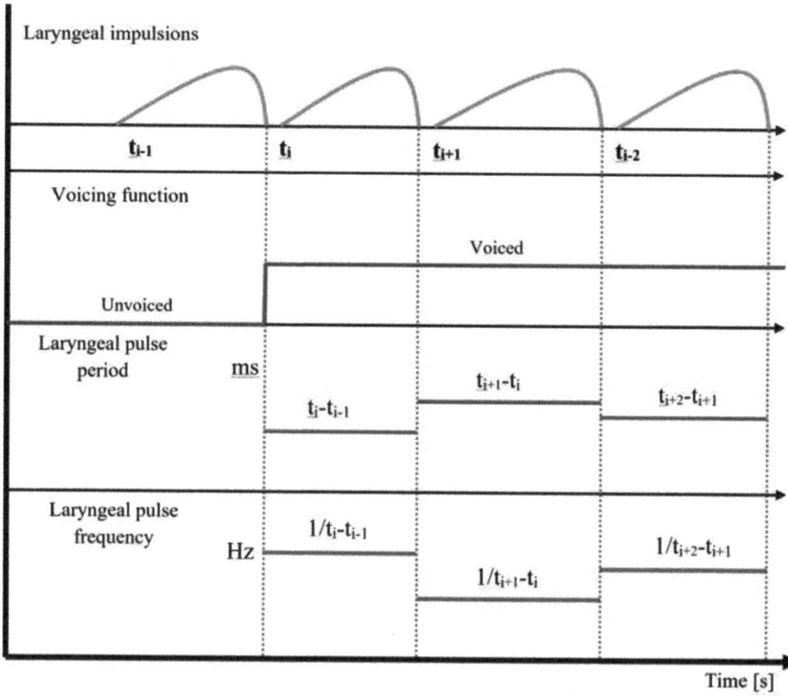

Figure 7.5. *Definition of laryngeal (pulse) frequency*

7.3. Laryngeal frequency and fundamental frequency

The so-called "voiced" speech sounds (vowels and consonants such as [b], [d], [g]) are produced with the vibration of the vocal folds, called laryngeal vibration. The laryngeal frequency, with symbol F_L, is measured directly from the vibration cycles of the vocal cords. The acoustic measurement of the fundamental frequency of the speech signal, symbol F0, is actually an estimate of the laryngeal frequency. F0, which is obtained from the acoustic signal, is therefore an estimate of F_L.

The laryngeal frequency can also be directly estimated by observing physiological data related to the vibration of the vocal cords (laryngograph). These physiological measurements identify the different phases of the glottic vibration cycle (laryngoscopy, variation of electrical impedance at the glottis, etc.) over time. In this case, if t_1 and t_2 designate the beginnings of two consecutive vibration cycles, the laryngeal period is equal to $T_L = t_2 - t_1$ and the laryngeal frequency is defined by (Figure 7.5):

$$F_L = 1/(t_2 - t_1) \text{ for } t_1 < t \leq t_2$$

The laryngeal frequency can vary considerably during phonation and can extend over several octaves. In extreme cases, transitions from 100 Hz to 300 Hz (change from normal phonation to falsetto mode) can be observed during an interval of two or three cycles.

Furthermore, successive cycles can show variations of several percent around a mean value, depending, among other things, on the physiological state of the muscles involved in the vibration mechanism. Even direct observation (for example, by rapid cinematography) does not always allow precise identification of the cycle beginnings, due to creaky voice, breath, etc. This may result in errors that are difficult to minimize.

The name "fundamental frequency", given to the acoustic measurement of laryngeal vibration, derives from the similarity with the same term given to the base frequency in a Fourier analysis, in other words, the presence of frequency harmonics integer multiples of the fundamental. This can sometimes result in confusion, which the context is not always sufficient to resolve.

F0 can be measured from the speech signal in the time domain, for example after signal filtering, or in the frequency domain, from the fundamental frequency (in the Fourier sense) of a voiced sound. The successive variations in F0 values over time are plotted in the graph to determine a so-called pitch curve, produced during phonation (Figure 7.6). This pitch curve conventionally displays null values at segments of unvoiced speech or silence.

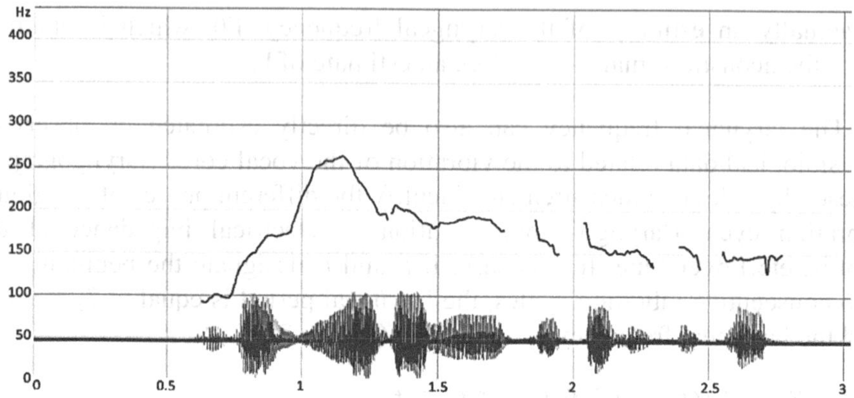

Figure 7.6. *Example of a pitch curve displayed as a function of time and varying from approximately 95 Hz to 260 Hz*

The difficulty in measuring the fundamental frequency is largely due to the fact that, strictly speaking, there are no glottic vibration cycles, but rather the recurrence of a movement that is controlled by numerous parameters (adductor and tension muscles controlling the vocal folds, pressure under the glottic, etc.). The speech signal which the measurement is made from is the result of the complex interaction of glottic stimulation and temporal variations in the shape of the vocal tract.

The mathematical tools usually used in signal processing and designed for the measurement of periodic phenomena, will often prove to be ill-suited to this kind of analysis, in which nothing is really stationary. For this reason, there are literally hundreds of procedures and algorithms for measuring speech fundamental frequency (Hess 1983), which can be classified according to their operating principles into temporal and frequency (or spectral) methods.

The difficulty in measuring F0 is due to several reasons:

– the fundamental component is sometimes absent in the signal, either because it has been filtered (in the case of old analog telephone landlines), or because it is very weak due to the acoustic nature of certain vowels (for example, in the case of [u] as in the French word *mou*);

– the presence of various noises in the signal (anything that does not result from the production of a single speech signal is considered as noise) makes it difficult to identify the harmonic components relevant to the calculation of F0 by a spectral method, and in particular that of the first component that is supposed to correspond to the fundamental. Voice overlap falls into this category;

– the coding and compression of the signal by various methods such as MP3, WMA, OGG, etc., generally introduces, after decompression, disturbances in the frequency values of the different harmonics, which can influence the measurement of F0, which is made from this information;

– phonation in creaky mode, for which the fundamental frequency is not clearly defined.

7.4. Temporal methods

Temporal methods attempt to measure successive laryngeal periods based on the evolution of the signal over time, without using spectral analysis. In principle, they avoid a windowing that would reduce the temporal resolution of the resulting pitch curve. When considering the patterns of laryngeal variation shown in Figure 7.7, the approach that seems the most natural is to identify a reference point in the cycle, such as the successive similar peaks, and to measure the time intervals, which should correspond to the (pseudo-) laryngeal periods.

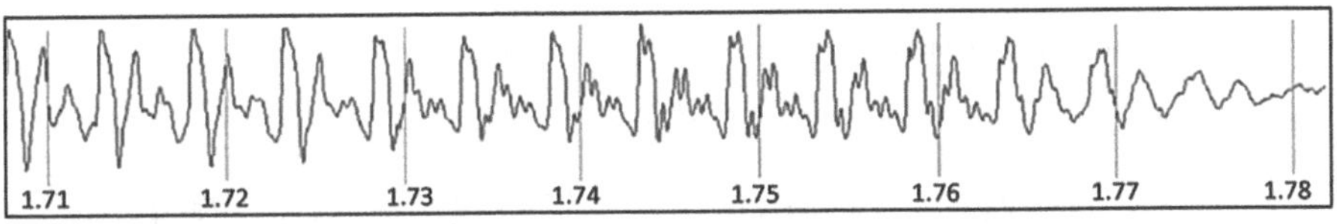

Figure 7.7. *Waveform (oscillographic curve) of a vowel [a]*

Although this task may visually appear to be easily accomplished, this is far from being the case for an electronic or algorithmic process. The difficulty lies in identifying the "right" reference moments, the "right" peaks, which is actually done visually by identifying the repeated patterns that make up a laryngeal cycle. Figures 7.7 and 7.8 illustrate this difficulty related to some harmonic phases varying from cycle to cycle: the "right" peak seems to move from one pattern to another, due to phase changes of several harmonic components.

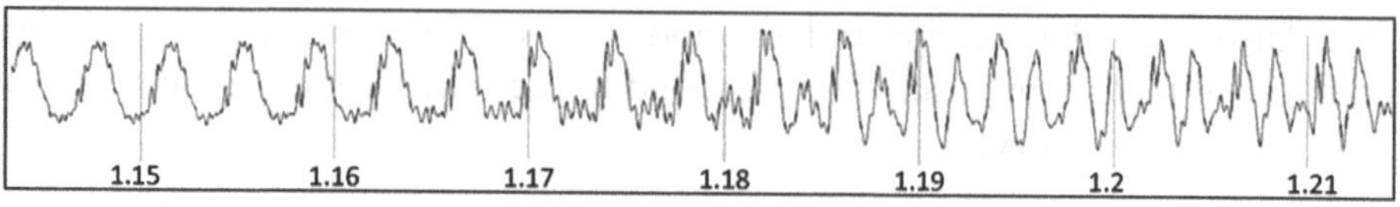

Figure 7.8. *Oscillographic curve of a vowel [i] showing the effect of harmonic phase shift*

In the early days of acoustic speech analysis, spectrographs were not available (Fourier harmonic analysis was calculated by hand), and the (approximate) calculation of the laryngeal frequency was based on the laryngeal vibration patterns obtained from the kymograph. After visually identifying the repetition of the characteristic patterns of each cycle, you could either make a direct measurement of the period by measuring the distance between two apparently similar instants of the vibration (top of Figure 7.9), or measure the duration of a number of periods (10 for example, as shown at the bottom of Figure 7.9). The latter method has the advantage of dividing the measurement error (which is inevitable, and caused by the thickness of the plot and the short length of the graphical interval among others) by the number of patterns considered, but at the price of obtaining a single frequency value for several laryngeal cycles, in other words, for a time window as in the case of spectral analysis.

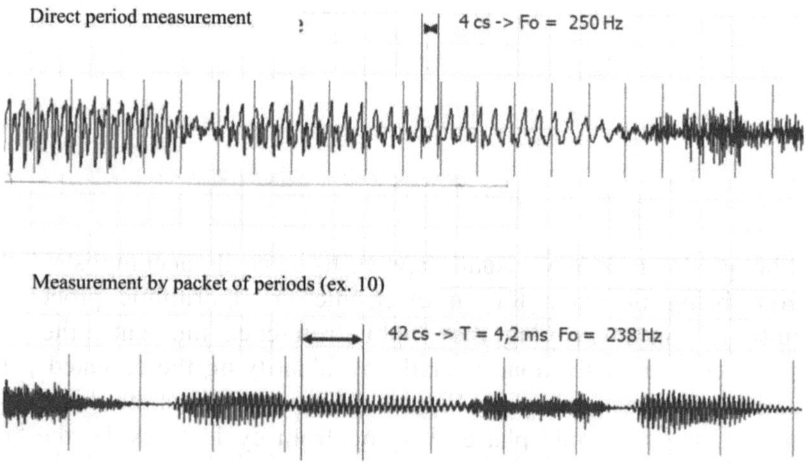

Figure 7.9. *Manual measurements of laryngeal frequency from the vowel waveform*

Since changes in the speech wave cycle patterns make the measurement difficult or imprecise, the ideal case of the sinusoid could be approached, for example, by using a low-pass filter, so that the fundamental harmonic component, alone, passes through the output, so the output signal only has one peak or two zero crossings per laryngeal period of the acoustic signal.

Then, two new problems arise. On one hand, the cut-off frequency of the low-pass filter will have to be adjusted so that the above condition is always met, regardless of the fundamental frequency value. However, this is generally an *a priori* unknown, or poorly known fact, since the fundamental is precisely what we are trying to measure. It will be necessary to manually, or automatically, adjust the filter or switch a bank of low-pass filters, in order to meet the measurement condition, a sine wave at the output, or at least no more than two zero crossings of the output signal per laryngeal period. Even automatic filter switching may be unsuitable in the case of rapid changes in the laryngeal frequency over several octaves. However, this technique has long been used by commercial pitchmeters (for example, Frokjaer-Jensen).

On the other hand, the measurement of the output signal is itself subject to error. If the periods are measured by detecting successive peaks, the error may be due to the presence of insufficiently filtered harmonics in the output signal and the resulting phase shift between the successive peaks. If it is done by detecting zero crossings of the output signal, the unavoidable presence of noise imposes a practical non-zero value of this "zero" value, and thus a phase shift, due to changes in the amplitude of the output signal (Figure 7.10). To minimize this type of error, compensation techniques by detecting positive and negative "zero" crossings have been used (Pitch Computer from (Martin 1975)).

Today, temporal methods are little used, unless they are driven by a more robust frequency method that frames the possible period values. The advantages of the frequency and time methods can then be combined by carrying out (pseudo) period-by-period measurements, which are necessary, for example, for the measurement of the jitter (cycle-to-cycle frequency variation) and the shimmer (cycle-to-cycle intensity variation).

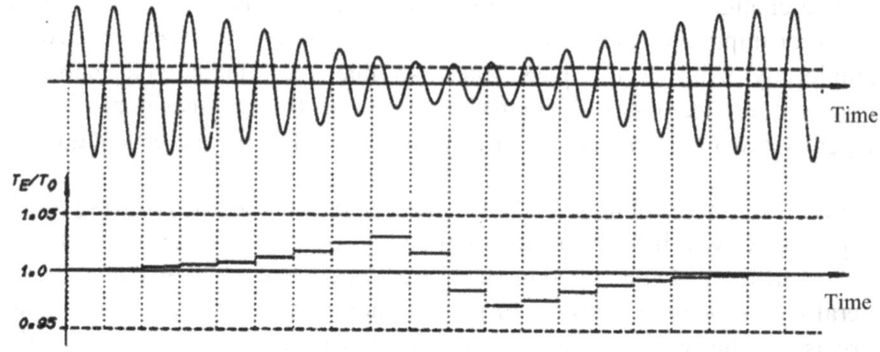

Figure 7.10. *Effect of amplitude variations on the measurement of periods by biased zero crossing*

7.4.1. *Filtering*

A simple device belonging to the category of temporal methods consists of recovering the fundamental frequency by means of a low-pass filter, with a cut-off frequency that can be adjusted manually or automatically according to variations in the fundamental frequency. The filter is followed by a frequency meter operating (for example) from the zero crossings of the filtered signal. The characteristics of the filter are chosen to eliminate unwanted harmonic components, so that the number of zero crossings of the filtered signal corresponds to that of the source. It can be shown (McKinney 1965) that this condition is fulfilled, if the sum of the amplitudes of the various harmonics that are greater than the fundamental multiplied by their harmonic rank, is less than the amplitude of the fundamental.

A low-pass filter with a fixed cut-off frequency is therefore suitable for speech sounds, where the production model predicts a sufficient fundamental amplitude with limited frequency variation, absence of occlusions, etc. The filter should be used for speech sounds that are not in a fixed cut-off frequency. Vowels such as [u], which often have a fundamental amplitude that is 10 dB to 12 dB less than that of the second harmonic, cannot be processed properly with this system and will require higher filter attenuation, resulting in the reduction of the useful analysis frequency band of the filter (Boë and Rakotofiringa 1971).

Furthermore, filtering inevitably causes phase shifts in the output harmonics, which can lead to errors in the automatic selection of filters configured in a filter bank. However, it is possible to compensate for these phase shifts by using delay lines (Léon and Martin 1969).

7.4.2. *Autocorrelation*

F0 measurement, by autocorrelation of a time window (from 10 ms to 50 ms duration), is a method that can now be implemented in real time, in addition to giving satisfactory results when the signal does not vary too much from one period to another (quasi-periodicity). The maximum of the autocorrelation function is obtained, in principle, when the offset between the original signal and the shifted signal is equal to one fundamental period (method used by the Praat software), thus simulating the visual identification of similar consecutive patterns of a voiced speech signal.

Unfortunately, for signals where the second harmonic is reinforced by the first formant, the maximum of the autocorrelation function corresponds to the second harmonic, giving an erroneous F0 measurement (frequency doubling), an error which can also occur during a visual inspection of the signal.

The mathematical formula for autocorrelation is:

$$R(\tau) = \sum_{-T/2}^{T/2} x_{n-\tau}\, x_n$$

which simply means that the value of this function for a given τ offset is obtained by summing the values of a window of duration T, multiplied by the values taken at the same place in another window of the signal offset by a duration τ. The maximum of this function corresponds to the fundamental period sought after (Figure 7.11).

It is therefore the duration of the time window that constitutes the parameter for calculating F0 (actually T0, the fundamental period) by autocorrelation. This duration must be greater than the expected fundamental period.

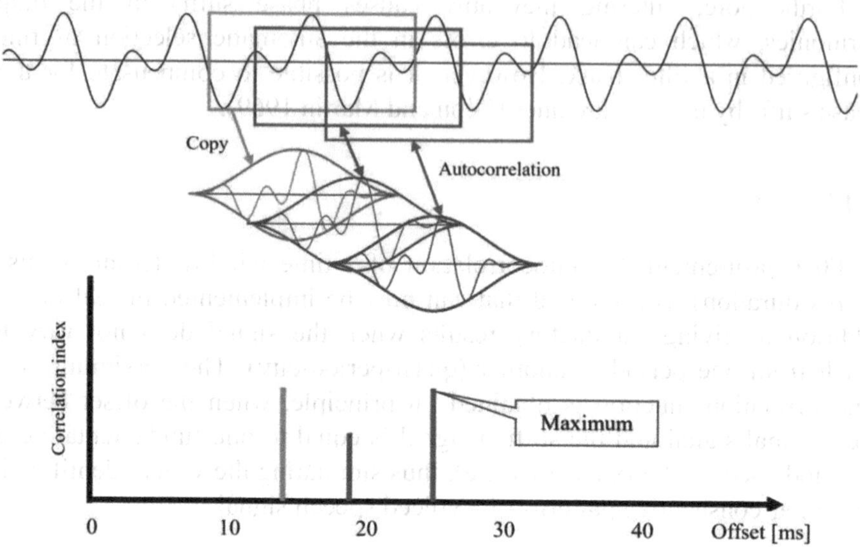

Figure 7.11. *Principle of F0 calculation by autocorrelation*

More or less heuristic non-linear pre-processing, such as peak clipping, center clipping, square or cube elevation, single or double rectification, sometimes improves the situation, by strengthening the amplitude of the fundamental, before calculating the autocorrelation, or even measuring zero crossings. Figure 7.12 shows two examples.

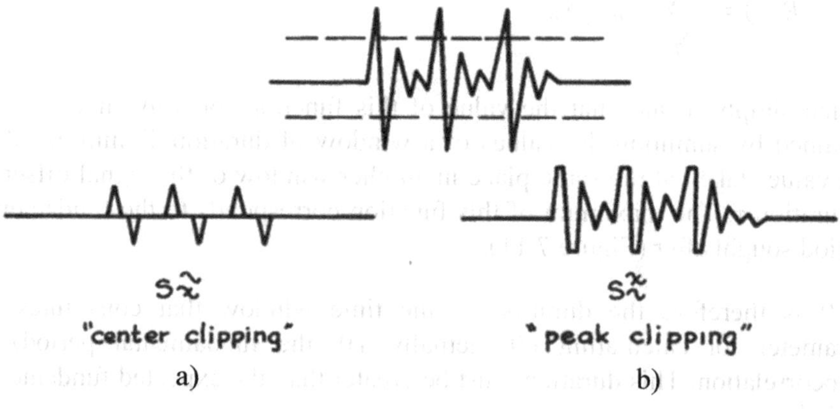

Figure 7.12. *Non-linear preprocessing of the signal center clipping (a) and peak clipping (b)*

7.4.3. *AMDF*

The temporal AMDF method (Average Magnitude Difference Function) was once widely used. This time, the aim is to find the minimum of a function within an expected period variation interval:

$$AMDF(\tau) = \frac{1}{N}\sum_{i=1}^{N}(x_i - x_{i+\tau})$$

which establishes which of the sums of absolute differences, taken term to term of the samples of two windows of duration T, gives the best match for offset τ (a perfect match gives a null AMDF value).

The AMDF method is therefore very similar to autocorrelation, with multiplications being replaced by differences, which was an advantage at a time when computer processors could only do addition and subtraction quickly.

7.5. Frequency (spectral) methods

Visual examination of a vowel spectrogram, or simply of a single spectrum that more or less is clearly revealing its harmonics, suggests that the use of frequency methods should be more effective than time methods. Indeed, since the fundamental frequency is equal to the frequency interval between two consecutive harmonics, it is theoretically sufficient to identify any two consecutive harmonics.

The acceleration of microprocessor clock frequencies in recent years has made it possible to use F0 analysis methods in the frequency domain, presupposing Fourier harmonic analysis as a pre-processing step. Instead of operating directly within a time window of the signal and determining its (pseudo) frequency, all or part of the harmonic information contained in the spectrum is used.

One might think that the simplest method is to use the first harmonic of the spectrum as the fundamental frequency. Unfortunately, the presence of noise of various kinds, and especially the possible absence of this first harmonic in the spectrum, makes this method unreliable at times. The signal delivered by analog (and sometimes digital) telephones is mostly devoid of components below 300 Hz (the bandwidth of an analog landline telephone signal, by design, ranges from 300 Hz to 3,400 Hz). However, the method of visually locating the fundamental from a narrowband spectrogram has long been used. In order to reduce the error in the value of F0, which is difficult to estimate because of the thickness of the harmonic curves, the frequency of the tenth harmonic was measured, and then divided by 10, in order to reduce the error in the value of the frequency accordingly (Figure 7.13).

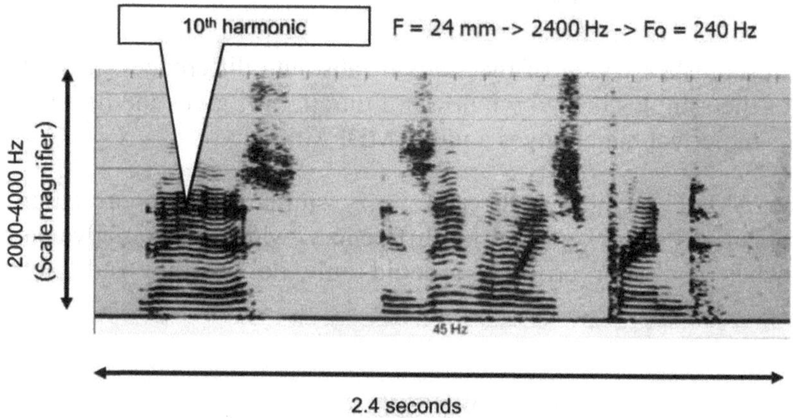

Figure 7.13. *Measurement of F0 from an analog narrowband spectrogram. To reduce the error, the frequency of the 10th harmonic is measured and then divided by 10*

If just the harmonic frequencies of the speech segment alone are available in a given spectrum, to the exclusion of other noise components, the fundamental frequency is obtained by assessing the largest common divider of the amplitude spectrum maxima. In practice, this implies that these maxima can be reliably identified, even in the presence of noise, and that a harmonic structure actually exists in the spectrum. Even if it corresponds to the fundamental, a spectrum with only one harmonic component is unsuitable.

7.5.1. *Cepstrum*

An old, now classic, method of this type of analysis is the cepstrum, which proceeds by the harmonic analysis of a signal spectrum (more precisely, of the logarithm of the signal amplitude spectrum). We can thus recognize the periodicity in the spectrum that is meant to correspond to the fundamental frequency desired (Figure 7.14).

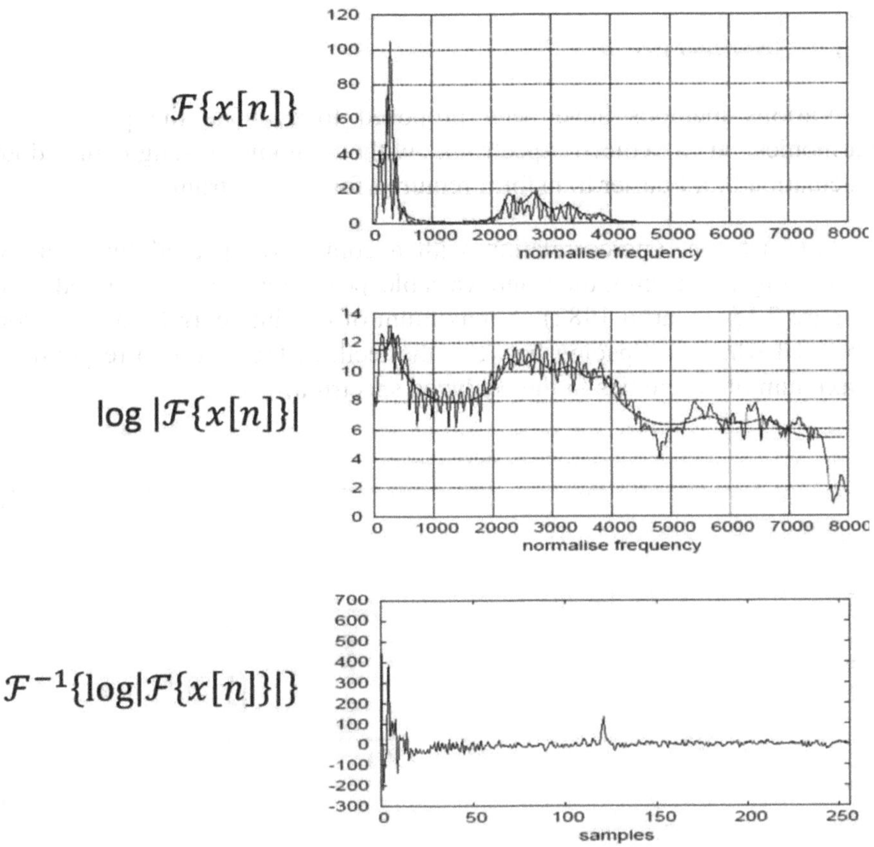

Figure 7.14. *Example calculation of a cepstrum: Fourier spectrum, logarithm of the Fourier spectrum and inverse Fourier spectrum of the logarithm of the Fourier spectrum (from (Taylor 2009))*

The amplitude of the cepstrum maximum is an indication of the degree of harmonicity of the components of the signal spectrum and is also an indication of the degree of voicing. A vowel will have a well-defined harmonic structure with respect to the noise components, and a large cepstral maximum.

On the contrary, an unvoiced consonant, therefore devoid of voicing and without regularly spaced harmonic peaks on the frequency scale, will give a cepstrum corresponding to the noise without a usable peak.

7.5.2. *Spectral comb*

Various methods have been proposed to evaluate the periodicity of harmonics in a voiced spectrum, without going through the double calculation of a Fourier transform required for the cepstrum.

For instance, intercorrelation with a comb-like spectral function, with decreasing tooth amplitude and variable peak spacing, gives good results (Figure 7.15) (Martin 1982). A maximum of this intercorrelation function is obtained when the spacing between the teeth of the comb corresponds to a maximum of harmonics in the analyzed spectrum.

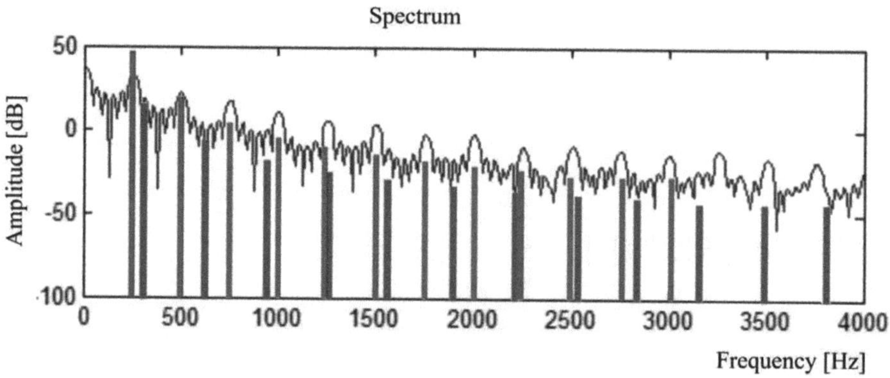

Figure 7.15. *F0 detection by the spectral comb method. Teeth of the optimal comb are displayed in red*

The presence of a harmonic structure, and the value of the harmonic interval corresponding to the fundamental frequency, are detected when the intercorrelation reaches a certain threshold, the value of which can also be used as a voicing criterion.

In general, spectral methods assume fewer constraints on the underlying source-filter model and are therefore more resistant to noise (in the sense defined above) than temporal analysis.

However, analysis by a spectral method requires a windowing of the signal, which implies an appropriate frequency resolution. For low F0 values, such as 70 Hz, a minimum frequency resolution in the order of 30 Hz is necessary, resulting in a sampling period of about 40 ms. Since the calculated value of F0 is relative to the entire window, fine or rapid changes in F0 cannot be measured directly. An analysis time of 32 ms, for example, will give a single F0 measurement, whereas at 300 Hz, more than nine estimated values of laryngeal frequency could theoretically be obtained. Therefore, a time window duration which is appropriate to the expected F0 values must be used each time, which is not always possible in the case of rapid voice register change.

The spectral methods, which are more resistant to noise, are suitable for the study of macro-variations of F0 (evolution of the pitch curve with respect to the syntactic structure, for example). Devices operating in the time domain, on the other hand, are desirable for the study of micro-melodies (cycle-to-cycle variations in phonation physiology).

7.5.3. *Spectral brush*

An extension of the comb method has been proposed by Martin (2000), under the name "spectral brush". The calculation of the maximum correlation value, with a comb function of decreasing amplitude, is no longer done on a single spectrum, but on all the spectra of a voiced segment. This method takes better account of the perception of the voicing, which is made over a certain duration of the signal and not over a single instant.

7.5.4. *SWIPE*

Swipe, which is a relatively recent method (Camacho 2007), is actually a variation of the spectral comb method, which does not use a comb function with decreasing tooth amplitudes with frequency, but a sawtooth function of variable frequency that has comb-like spectral characteristics (Figure 7.16). A selection of the first and second harmonics (one of the optimization parameters of the comb method) partly eliminates the erroneous detection of sub-harmonics.

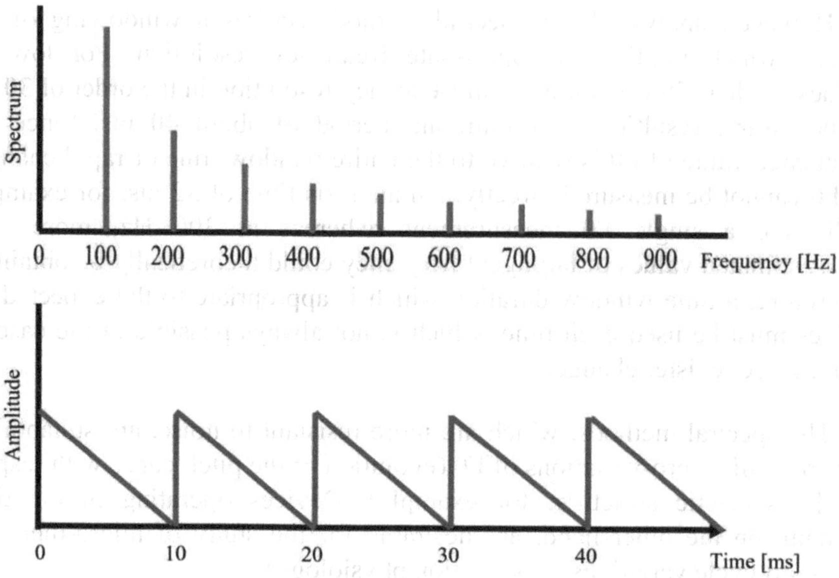

Figure 7.16. *The sawtooth function and its spectrum used by SWIPE to detect F0*

7.5.5. *Measuring errors of F0*

Despite their complexity and the ingenuity of the algorithms, all devices developed to date have failed under specific conditions. These conditions can be determined, to a certain extent, by means of the model involved in the principle of analysis. The errors can be divided into two groups:

1) so-called "gross" errors, for which the value obtained deviates considerably (by more than 50%, for example) from the "theoretical" fundamental. This is the case for harmonic identification errors, where the analyzer proposes a value corresponding to the second or third harmonic, and "misses" due to a temporary drop in the amplitude of the filtered fundamental (the case of the time domain);

2) so-called "fine" errors, for which the difference in F0 measured in relation to the laryngeal frequency, measured cycle by cycle, is only a few percent. The fine errors are mainly due to the interaction of the noise components when the amplitude of the fundamental is low.

In practice, despite the emergence of increasingly sophisticated processes, reliable detection of the fundamental frequency requires a good quality speech signal (frequency response of the recording and absence of noise) and, in most cases, the actual presence of the fundamental component in the signal. In practice, it is advisable to always display a narrowband spectrogram to visually check the relevance of the F0 curve display, and to correct the analysis parameters if necessary.

7.6. Smoothing

The successive measures of F0, by temporal or frequency methods, sometimes present confusing aspects when they appear in the form of a pitch curve. These curves can present many irregularities and errors, especially during transitions from voiced to unvoiced modes, and vice versa. Most software uses a smoothing process in order to obtain a curve that appears more regular and more "pleasant" to the observer. Mathematical smoothing methods generally use a dynamic programming algorithm: the pitch curve results from the calculation of an optimum path that is found from the successive "raw" values of F0, replacing atypical values, if necessary – therefore classified as erroneous by the algorithm – relative to values close in time, by values that seem more likely according to a certain criterion.

A median filter smoothing is added to this type of correction, which replaces a given value with the median of a table of a certain number of consecutive F0 values (usually a median filter of the 3rd or 4th degree, using a table of seven or nine consecutive F0 values). Through the use of such a filter, the selected value of F0 can result from a disruption in the temporal order of the raw values of F0.

Lastly, a Gaussian smoothing often completes these first two operations, giving the Gaussian curve-weighted average of a certain number of successive values, which is delivered by the previous operation as the final value of F0.

While the final result of these smoothing operations may be more satisfactory from the perspective of the graphical regularity of the pitch curve, one should not lose sight of the fact that these "corrections" introduce artefacts into the final pitch curve, the most remarkable effect of which, is to mask the fine variations in the pitch curve. Figures 7.17a and 7.17b give an example of a pitch curve with and without smoothing (dynamically programmed smoothing, median filter of order 3 and Gaussian average).

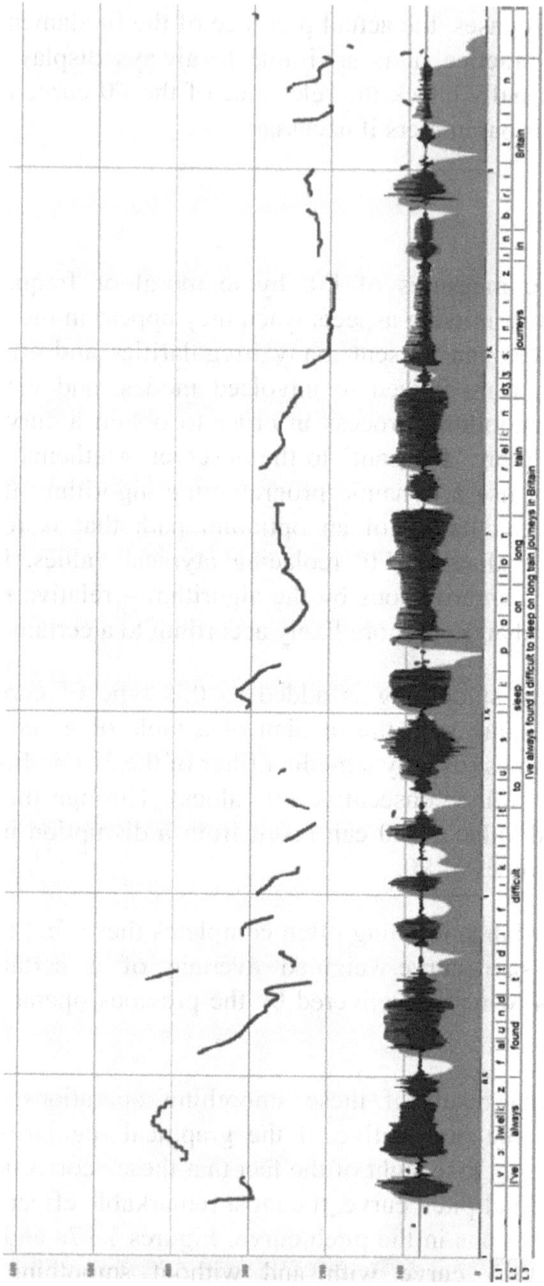

Figure 7.17a. *Pitch curve of the sentence "I've always found it difficult to sleep on long train journeys in Britain" without smoothing (Anglish corpus)*

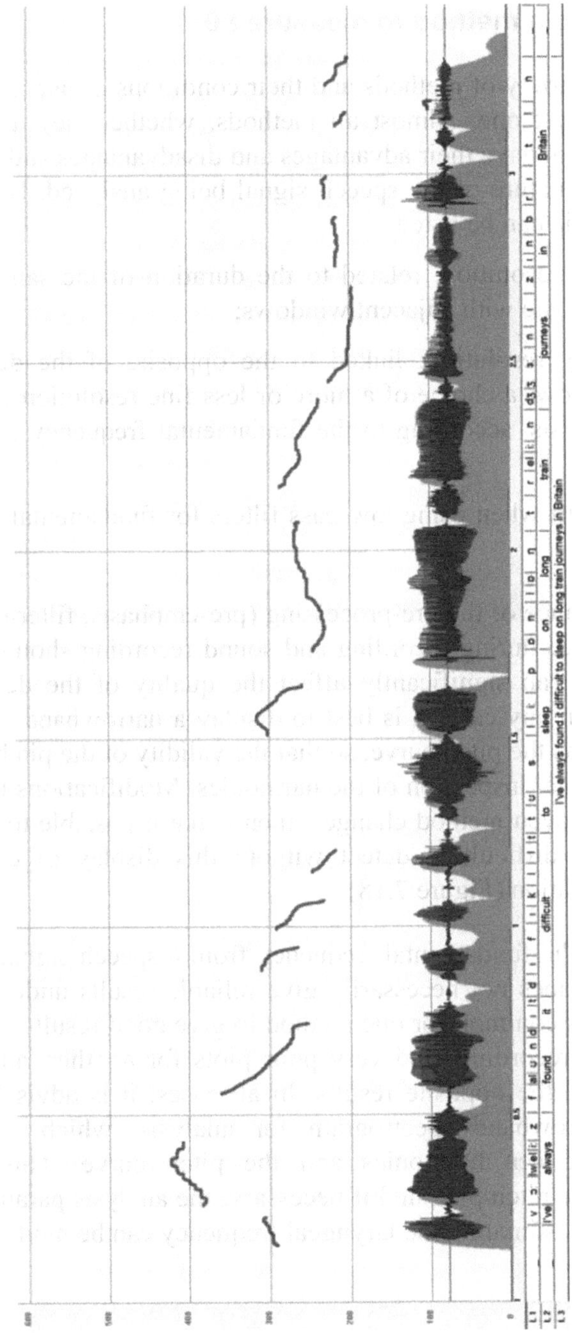

Figure 7.17b. *Pitch curve of the sentence "I've always found it difficult to sleep on long train journeys in Britain" with smoothing by dynamic programming (Anglish corpus) (continued)*

7.7. Choosing a method to measure F0

The wide variety of methods and their conditions of use make it difficult to select the best one. Almost all methods, whether they are temporal or frequency-based, have their advantages and disadvantages and their use must depend on the nature of the speech signal being analyzed. However, a few selection criteria can be listed:

1) temporal resolution: related to the duration of the sampling window and its overlap rate with adjacent windows;

2) frequency resolution: linked to the opposite of the duration of the sampling window; a choice of a more or less fine resolution for the Fourier harmonic analysis, according to the fundamental frequency range that is to be measured;

3) phase shift when using low-pass filters for fundamental overlap in the time domain.

The importance of the pre-processing (pre-emphasis, filtering) that would have taken place during recording and sound recording should also be kept in mind, as it can significantly affect the quality of the detection of the fundamental. In any case, it is best to display a narrowband spectrogram at the same time as the pitch curve, so that the validity of the pitch curve can be checked by visual inspection of the harmonics. Modifications to the analysis parameters, or even method changes, then make it possible to correct errors that would be difficult to detect without this display of complementary spectral information (Figure 7.18).

Measuring the fundamental frequency from a speech signal is a complex operation that does not necessarily give reliable results under all recording conditions. It is common for one method to give good results for a particular section of the recording, and very poor plots for another, and for another method to give the opposite results. In all cases, it is advisable to always display a narrowband spectrogram for analysis, which allows you to distinguish between harmonics and the pitch curve. This ensures the reliability of the pitch plot, and if necessary, the analysis parameters or even the method for estimating the laryngeal frequency can be modified.

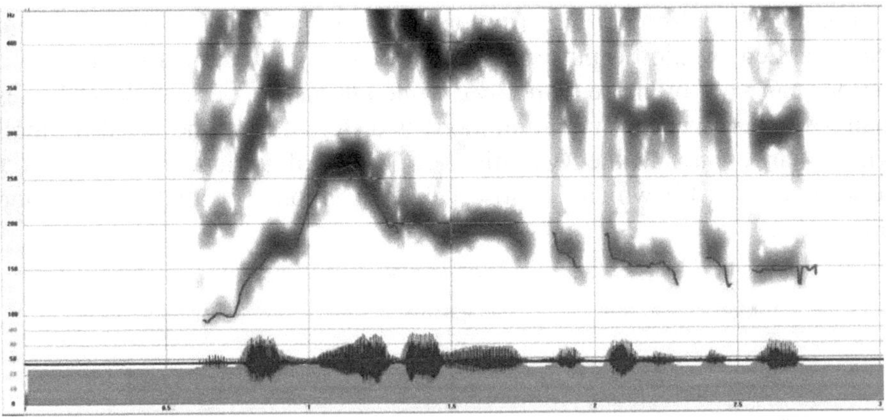

Figure 7.18. *Simultaneous display of the pitch curve and a narrowband spectrogram, allowing visual inspection to identify possible measurement errors of F0*

7.8. Creaky voice

There are several types of voices perceived as *creaky* that can be characterized in a speech signal, either by great irregularity in the pitch periods (vocal fry, Figure 7.19), by a very low fundamental frequency F0 (< 50 Hz), or by an impulse pairing (Figure 7.20).

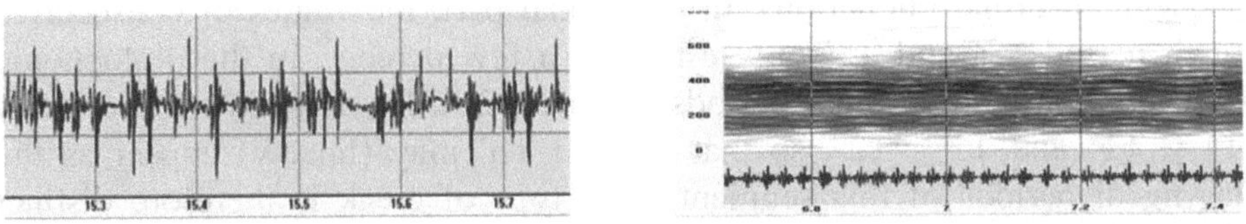

Figure 7.19. *Creaky segment with irregular pitch periods (vocal fry)*

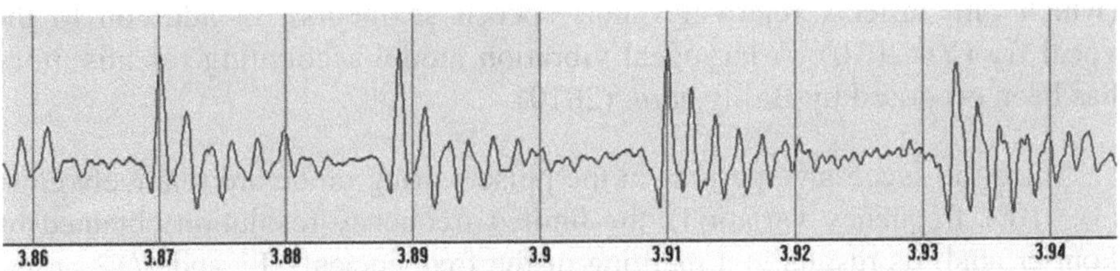

Figure 7.20. *Creaky segment with very long laryngeal periods*

Laryngeal pulse pairing (diplophonia) is a type of creaky voice characterized by glottic pulse pairing (McKinney 1965): consecutive pitch periods can vary by more than 10%, while alternating periods differ by less than 5%. This laryngeal vibration mode is different from the vocal fry mode with a large jitter value (variation of laryngeal period, from period to period), a mode for which no close values are found for alternating periods (Figure 7.19).

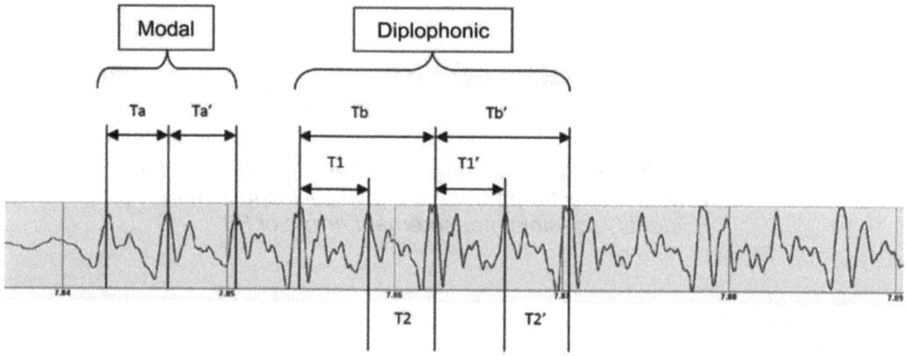

Figure 7.21. *Creaky segment with laryngeal period pairing (pitch doubling)*

Figure 7.21 shows a case of diplophonia, that appears following a modal sequence (normal phonation). In the modal part, the values of consecutive pitch periods Ta and Ta' only differ by a few percent. In the diplophonic part, the consecutive pitch periods T1 and T2 differ more in duration than T1 – T1' and T2 – T2', so that Tb and Tb' may (falsely) appear as the laryngeal periods of this segment. This type of creak is therefore distinct from other modes of laryngeal vibration, which are irregular (Figure 7.19) and have very low frequency (50 Hz, Figure 7.20). The laryngealization observed in the production of Tone 3 in Mandarin also uses this mode (which only affects relatively short speech segments), in addition to the vocal fry (Yu 2010). A laryngeal vibration model accounting for this mode has been proposed by Bailly *et al.* (2010).

Since the successive periods in the pulse pairing mode are relatively close (+/– 10% frequency variation), the limited frequency resolution obtained by Fourier analysis results in a merging of the frequencies 1/T1 and 1/T2 on the spectrograms. However, since a temporal regularity is created with a period Tb (with Tb = T1 + T2) this time, a new harmonic component (sometimes called sub-harmonic) appears with a frequency 1/Tb (Figure 7.22).

Figure 7.22 shows an example of a modal-diplophonic-modal sequence. The diplophonic segment displays the "sub-harmonics" with an initial component of very low intensity.

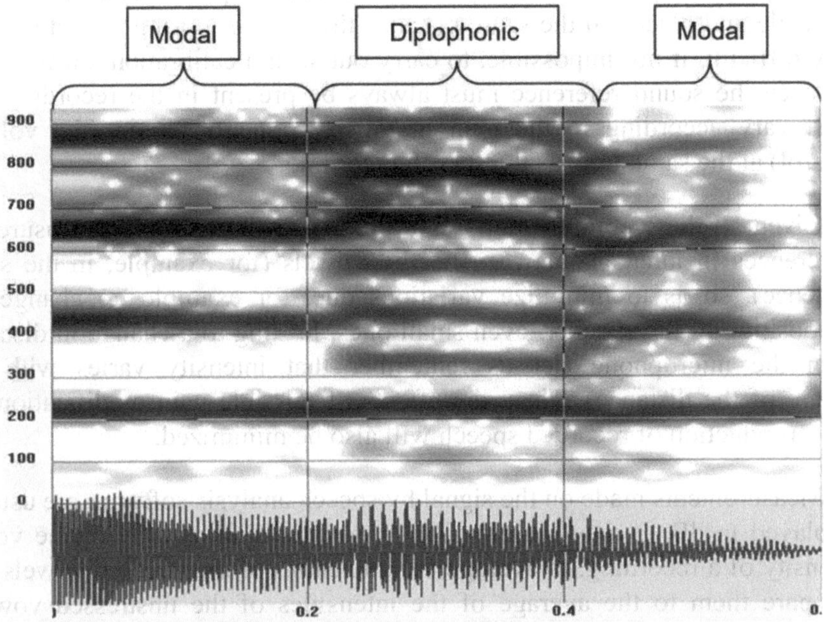

Figure 7.22. *Spectrogram of a modal-diplophonic-modal sequence*

This is a consequence of the time pattern of the creaky segment, with a low amplitude of its fundamental component. Vowels, such as [u] in French, sometimes have the same characteristic with an initial harmonic component that has an amplitude that is 6 dB to 12 dB lower than that of the second harmonic.

There is also a shift in the frequencies of "modal" harmonics, due to the confusion caused by the Fourier analysis of two harmonic peaks of neighboring frequencies.

7.9. Intensity measurement

We have seen that intensity is proportional to the square of the amplitude of a pure sound, and is usually measured in decibels relative to a reference amplitude or intensity that is equal to 0 dB by definition. The absolute intensity of a vowel, for example, can therefore only be measured in relation to this 0 dB reference. However, unless a calibrated pure sound generator is available and placed in the same place as the speaker during recording, it is very difficult, if not impossible, to carry out such a calibration, all the more so since the sound reference must always be present in the recording and must vary according to the gain in the reproduction chain (the volume control) at the same time as the speech recording.

Given all these complications, it is preferable in practice to measure the differences in intensity between nearby vowels (for example, in the same sentence), so as to minimize variations due, for example, to changes in the position of the speaker, even small ones, leading to a change in distance from the microphone (let us remember that intensity varies with the square of the distance). Thus, the effects of possible gain modifications in the reproduction of recorded speech will also be minimized.

Measurements made on the signal by speech analysis software are usually displayed in dB. It may be tempting, for example in a study of the vowel intensity of a recording, to average the intensities of the stressed vowels and compare them to the average of the intensities of the unstressed vowels. However, averaging intensity (or amplitude) in dB introduces an error, since the logarithm of a sum is not equal to the sum of the logarithms of the summed elements. Moreover, it is the amplitudes that add up, not the squares of the amplitudes relative to the intensity.

Let us assume we have two intensity values, 20 dB and 60 dB. The average in dB is therefore 40 dB. However, 40 dB corresponds to an intensity I_1 of 20 log(I_1) = 20, so $I_1 = 10^1 = 10$, whereas if I_2 has an intensity of 60 dB, $I_2 = 10^3 = 1{,}000$. The corresponding amplitudes are $A_1 = \sqrt{10} = 3{,}16$ and $A_2 = \sqrt{1\,000} = 31{,}6$. The average of the amplitudes is therefore (3.16 + 31.6)/2 = 17.38, which corresponds to an average intensity of 24.4 dB, which is quite different from the average intensity of 40 dB announced at the beginning.

What if you want to make perceived loudness measurements using average audiometric curves? It is possible, and this has been implemented in specialized measuring devices to decompose the signal corresponding, for example, to a vowel section by means of a Fourier analysis. Each component resulting from this analysis corresponds to a pure sound (under the analysis conditions described above), and you can therefore, theoretically, use the ear's response curves to obtain perceived dB values from the absolute dB of each component. The problem is that, in practice, only relative values, not absolute values of intensities, are known. It will therefore be necessary to make an approximation based on an average of the existing dynamic range compression for low frequencies below 500 Hz, and high frequencies above about 7,000 Hz.

If we limit ourselves to a frequency range of 500 Hz to 4,000 Hz, which covers most frequencies of speech sounds, except the fundamental and first harmonics, a reasonable approximation seems to be possible, the maximum error being about 5 dB, which can be compensated for by the enhancement of frequencies above 1,000 Hz, carried out by default by most spectrographic software packages.

The use of a Hamming, Hann(ing), or other sampling windows also minimizes echo effects, due to overlapping and the addition of phase-shifted harmonic components.

7.10. Prosodic annotation

The orthographic transcription of a speech recording is a prerequisite for all kinds of annotations, be it morphological, syntactic, informational, but very rarely prosodic.

In fact, prosodic transcription, which is carried out through punctuation, only very partially represents the intonation of the sentence. As for prosodic information related to attitudes, emotions or socio-geographical particularities of the speakers, it is often described by global parameters involving, for example, the mean and standard deviation of the fundamental frequency and syllabic durations.

Consequently, prosodic annotations describing objects in the linguistic system focus on pitch events localized on syllables, and in particular for

French, on stressed syllables, and also for English, on syntactic group final syllables. However, we speak and read by splitting the flow of speech into groups of words, by a segmentation called phrasing (by analogy with musical phrasing). Phrasing groups words together so that they only contain one stressed syllable (excluding emphasis stress), hence the denomination *accent phrases*. However, accent phrases are not simply chained one after the other. They can themselves be grouped into larger units, often called *intermediate* (ip) and *Intonation phrases* (IP), similar, but not necessarily congruent, to the syntactic structure. These groupings in successive levels, which organize and group accent phrases, constitute the prosodic structure.

7.10.1. *Composition of accent phrases*

The perception of speech is achieved through the segmentation into accent phrases, and their hierarchical grouping into a prosodic structure. This means that the apprehension of a text during oral or silent reading, and also the decoding of read or spontaneous speech, is done through the use of phrasing and the prosodic structure.

This structuring will influence our access to the meaning of a text and is inevitable. It is impossible for us to speak or read a text, even silently, without (re)generating a prosodic structure (except for very rare cases of readers proceeding solely from the graphic images of words and not from their sound images).

For languages with a lexical stress, such as English or Italian, accent phrases usually consist of just one lexical word, verb, adverb, noun or adjective, together with grammatical words such as pronouns, conjunctions, prepositions, articles, etc. This is not at all the case for languages without lexical stress, such as French.

For French, but also for Korean and other languages without lexical stress, the composition of an accent phrase depends on the read or spontaneous speech rate, whether oral or silent.

Therefore, for these languages, segmentation into accent phrases is not unambiguous, since it depends on the speech rate achieved by the speaker, or the reading speed adopted by the reader. A fast speech rate or speed favors longer groups, and conversely, a slow speech rate or reading speed leads to segmentation into shorter groups, each with only one syllable, which is necessarily stressed.

This implies that a single accent phrase may contain more than one lexical word, such as (*la ville de MEAUx*) "the city of Meaux" pronounced orally or in silent speech with a high syllabic rate (6 or 7 syllables per second, for example), whereas a slower pronunciation or reading (4 syllables/second) results in a phrasing comprising of two accent phrases (*la v***I***lle*) and (*de MEAUx*). By slowing down the rate of flow even further, we obtain single-syllable stress groups: (*l***A**), (*v***I***lle*), (*d***E**), (*MEAUx*).

7.10.2. *Annotation of stressed syllables, mission impossible?*

The first step in prosodic annotation is the identification of stressed syllables. The mere fact that we know (or think we know) the possible positions of stressed syllables when we read aloud or silently, suggests that we do not really need acoustical data to perceive stressed (non-emphatic) syllables when annotating a corpus of speech. Not only can reading the same text aloud or silently lead to different segmentations, but when listening, one cannot help but have expectations of a potentially different location of stressed syllables, than the ones produced by the speaker. In other words, we may perceive syllables that may not be acoustically stressed as being stressed, and conversely, we may not perceive syllables that have the expected characteristics of duration and pitch variation of stressed syllables as being stressed. This apparent illusion is found in many processes involved in speech perception (Arnal and Giraud 2017), and does not involve the direct processing of an acoustic input, but the validation of an expected input by a comparison between what is expected and what is physically realized.

Given that the identification of accent phrases depends on the syllabic rate chosen by the speaker or reader, stress annotation may be influenced by the rate at which the annotators are speaking, not in their speech, but in their listening. They will then be able to note syllables that they themselves would accentuate in the same context as accentuated, even though these syllables would not present characteristics of duration and/or pitch variations that would distinguish them from the surrounding syllables. Conversely, some annotators will not pick up the accentuation of syllables in cases where they would not accentuate them themselves, whereas these same syllables would have the acoustic characteristics of stress.

The problem for an annotator of stressed syllables, especially in French, is therefore to adapt to the speech rate of the recording. Perception will be influenced by the annotator's prediction process, which tends to detect stressed syllables where it would have placed them by reading or speaking – not with the speaker's rate, but with its own. These difficulties have been observed (but not explained) many times in work on the annotation of stressed syllables on corpora of spontaneous speech (Avanzi 2013; Christodoulides and Avanzi 2014).

7.10.3. *Framing the annotation of stressed syllables*

This dependence on the rate of reading or perception obviously raises an issue regarding the reliability of prosodic annotation, which primarily depends on the phrasing, in other words, the annotation of stressed syllables. However, if the segmentation noted by the annotator depends on the syllabic rate, there is a limit – not on the number of syllables that an accent phrase may contain, but on the interval of enunciation duration between two successive stressed syllables in continuous speech (and therefore also in reading speech), which cannot be less than 250 ms, nor longer than about 1,250/1,350 ms (Martin 2018b). As a result, an accent phrase cannot have a reading or pronunciation duration longer than 1,250/1,350 ms. The existence of this upper limit can easily be realized by silently or orally pronouncing words that are orthographically very long, such as *antidisestablishment-Arianism* or *paraskevidekatriaphobIa*, which require a so-called secondary, non-primary stress, even in silent reading.

As for the lower limit of 250 ms, it can be easily established by measuring the time taken for the fastest possible silent reading of a text and dividing this time by the number of accent phrases, a number which is roughly estimated since it depends on the reading speed. These time limits on the duration of accent phrases, apparently related to the ranges of time variation of delta brain oscillations (Martin 2018b), can be used as a framework for the first stage of prosodic annotation, by drawing the annotator's attention to intervals that are too large, or too short, between two successive stressed syllable notations.

7.10.4. *Pitch accent*

The second stage of prosodic annotation involves the annotation of pitch variations in the place of syllables, or more precisely, stressed vowels. The syllabic stress in the final position of the accent phrase is not a stress of intensity, contrary to a belief that is still widespread among some phonologists specializing in ancient languages. It is a stress that manifests itself through a combination of syllabic duration and pitch variation, essentially carried by the vowel of the stressed syllable. The voiced consonant that follows this vowel can, however, participate in the perception of the stress, but is not a determining factor in the mechanism of indicating the boundaries of the accent phrases, or in the prosodic structure.

A well-trained musical ear can easily determine degrees of stress among the stressed syllables of a phrase, and even discern the ups, downs or pitch flats. An approximation of this expertise can be made by a relatively simple calculation on the pitch curve, instantiated in the acoustic analysis of speech by the fundamental frequency. In the best case, that is to say, when the recordings are of good quality, both in terms of the range of frequencies reproduced and the low background noise level, the fundamental frequency is a relatively accurate measurement of the vibration frequency of the vocal folds.

The fundamental frequency F0 typically evolves, as we have seen, in the range of 70 Hz to 200 Hz on average for a male voice, and from 150 Hz to 300 Hz for a female voice. At the location of the stressed vowels, the prosodic variation can take various forms, which can usually be approached linearly by a single line segment, which connects the points on the fundamental frequency curve corresponding to the beginning and end of the stressed vowel. The details of the pitch variation, appearing in a slightly concave or convex form, result from particular regional or idiosyncratic realizations (Martin 2009).

The linear approximation of the variation of F0 at the location of the stressed vowels, in the final position of each accent phrase, thus describes the speed of pitch variation, which can be characterized by its glissando value. This value, which is obtained by dividing the frequency difference in semitones by the duration of the contour, is characteristic with respect to a glissando threshold value, above which the pitch variation is perceived as such (in other words, as a change in frequency), and below which it is perceived by a static tone (placed at 2/3 of the pitch variation). This is an approximation

(Rossi 1971), with the threshold being obtained by dividing a coefficient that typically varies from 0.16 to 0.32 by the square of the duration of the contour.

7.10.5. *Tonal targets and pitch contours*

The autosegmental-metric model is dominant in the field of linguistic descriptions of sentence intonation. Prosodic events, by hypothesis located on the *a priori* remarkable points of the accent phrases, in other words, the first, the stressed and the final syllables, are described by high and low abstract tonal targets, using the symbols H and L embellished with various clues indicating their positioning (%H/L and H/L% at the beginning or end of the prosodic phrase boundary, H* and L* aligned with an stressed syllable, etc.). The problem with this "ToBI" notation (acronym for "Tone and Break Indices" (for English, see Beckman *et al.* 2005, for French, see Delais *et al.* 2015)) comes from the relative vagueness linked to the interpretation of what can be a high or low tonal target, leading, in practice, to interpreting even minimal, and therefore unperceived, variations as tonal targets in acoustic pitch curves.

In contrast, notation by pitch contours, categorized by their glissando value, makes it possible to integrate the perceptive dimension of pitch variations. Indeed, albeit approximate, the calculation of the glissando makes it possible to classify the contours according to their value that is above or below a threshold of perception of pitch variation, and thus to integrate their assumed function into the prosodic notation, in other words, to indicate relationships of dependency that exist between accent phrases, in order to form the prosodic structure.

7.11. Prosodic morphing

Prosodic morphing consists of modifying the source parameters, without altering the properties of the filter constituted by the vocal tract. In phonetic terms, the fundamental frequency, intensity and duration parameters are therefore modified, with the aim of disrupting the segmental characteristics of vowels and consonants as little as possible. To this end, two methods are often used: Psola, operating in the time domain, and the phase vocoder, operating in the frequency domain. This process belongs to the domain of experimental phonetics, characterized by modifications to the speech signal, as opposed to instrumental phonetics, which analyzes the data without modifying it.

7.11.1. *Change in intensity*

The change in intensity is trivial: simply multiply each sample of the signal by an appropriate coefficient. The only difficulty is avoiding the digital saturation of the system, which is conditioned by the conversion format (8 bits, 12 bits, etc.).

7.11.2. *Change in duration by the Psola method*

The change in duration had already been carried out by analog means in the 1970s. A tape recorder with several rotating playback heads was used to play back time segments of the reproduced signal, during the slow motion of the magnetic tape, and thus produce a slow-motion effect on the speech rate with acceptable distortions. This system therefore proceeds by copying speech (or music) segments that are reinserted at a regular speed in time.

This principle was taken up again in the Psola method (Pitch Synchronous Overlap and Add (Moulines *et al.* 1989)), but this time, by extracting segments taken with an appropriate window (Hann(ing), for example), synchronized with the peaks of the laryngeal periods (Figure 7.23).

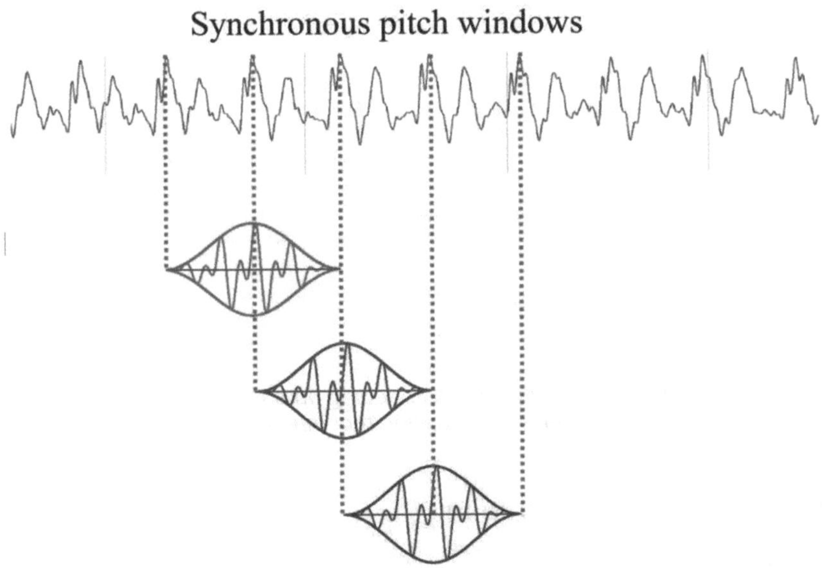

Figure 7.23. *Principle of speech signal decomposition by the Psola method*

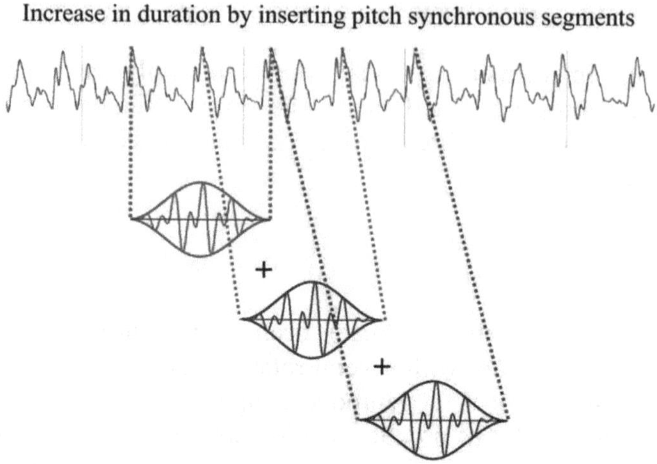

Figure 7.24. *Slowing and speeding up speech through Psola decomposition/recomposition*

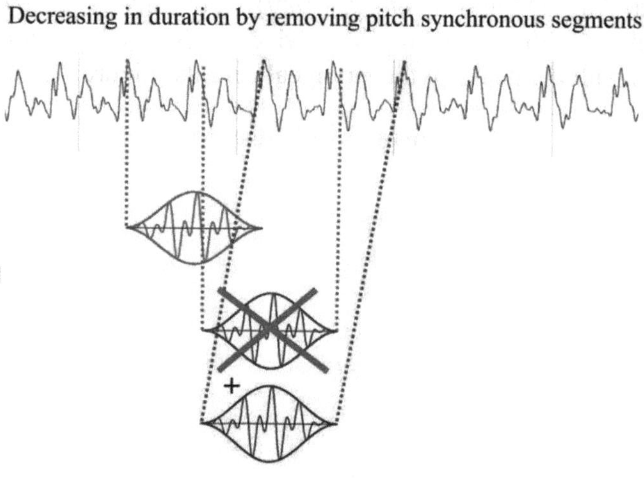

Figure 7.25. *Slowing and speeding up speech through Psola decomposition/recomposition*

7.11.3. *Slowdown/acceleration*

To slow down the signal, a copy of each segment taken at a regular rhythm is inserted: one insertion every three pitch periods produces a slow down of ¼, in other words, 25%. To speed up the signal by 25%, for example, we delete one sample out of four and connect the remaining segments by superimposing them. The disadvantage of this method is that the laryngeal periods for the adjacent parts have to be marked, in order to determine the width of the sampling window used and to assemble the segments. If the duration of the sampled segments is too long, there will be more overlapping of the assembled segments and the possibility of an echo, since different phase harmonics will most often be added together. If the duration of the segments is too short, there can be a loss on the low frequency side and especially on the fundamental.

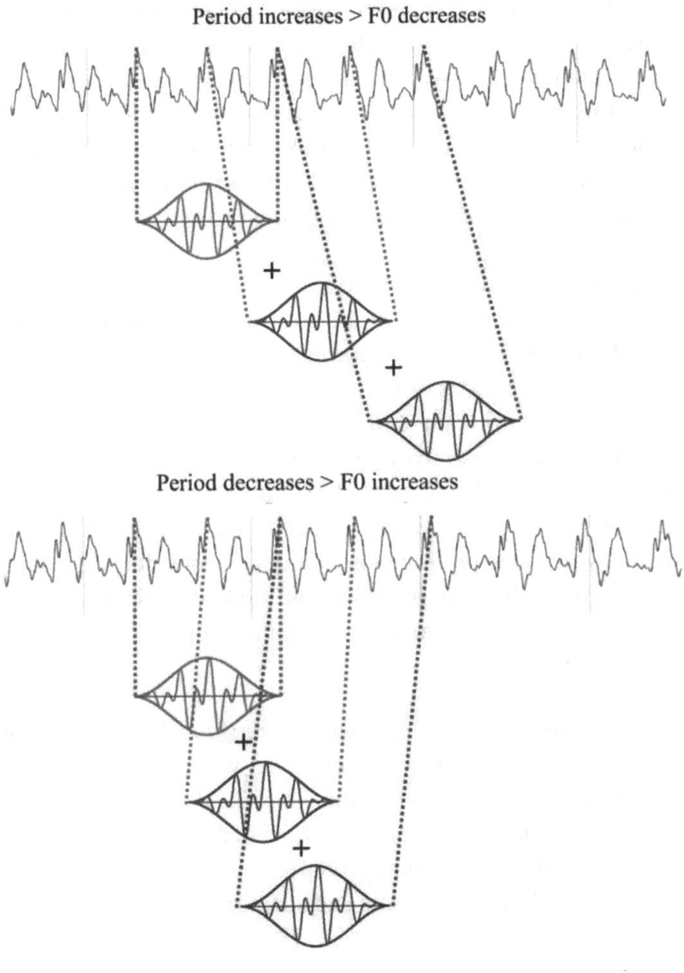

Figure 7.26. *Modification of the fundamental frequency of speech through Psola decomposition/recomposition*

A window with a fixed duration, such as 30 ms, is used for the unvoiced parts. Transients, due to occlusion, should be detected in order to avoid duplication when slowing down, or suppression when speeding up. However, this is not really necessary in practice, as possible distortions are generally acceptable in speech perception research.

7.11.4. *F0 modification*

To change the fundamental frequency of the signal by the Psola method, the segments taken at each laryngeal period are assembled by moving them closer together to achieve a shorter period – thus obtaining a higher fundamental frequency – or by moving them apart on the time axis in order to achieve a longer period and thus a lower frequency (Figure 7.25). However, when the overlap of contiguous segments is too great, a limit is reached to increase the frequency, producing an addition of harmonics that are out of phase with each other, resulting in a more or less important echo.

One wonders how a process as simple as Psola can give such satisfying results, which were previously only obtained at the cost of heavy calculations related to the phase vocoder. If the slowing down and acceleration of the speech rate by copying or erasing n periods by m periods of the signal is intuitively easy to understand, how can we explain that the formant structure is (relatively) preserved by copying successive overlapping windows (case of an increase in F0), or by spacing them out by leaving a short silence between each window (case of a decrease in F0)? Incidentally, Christian Hamon, a young trainee at the *Centre national d'études des télécommunications de Lannion* (Lannion National Center for Telecommunications Studies), the first (probably) to propose this method, left the engineers of the "speech analysis" research team at the time rather doubtful as to the results that could be expected from it.

In fact, as Kanru Hua noted in his blog *reunderstand Psola* (2015), it all comes down to the choice of time windows, in other words, their duration and their synchronization with the laryngeal pulse moments. The duration of the window of less than two times the period and its alignment with the pitch peaks guarantees the preservation of the wideband spectrum, given that it is limited to two periods. Kanru Hua gives the following details (translated and adapted):

"Psola first implicitly estimates a wideband spectrum array from the input signal, with each instant of analysis synchronized over the period; then it implicitly reconstructs the narrowband spectrum by subsampling the wideband spectrum with a modified fundamental frequency. With no pitch modification (in other words, modification of the time scale alone), the narrowband spectrum can be perfectly preserved throughout the process.

In the context of the source-filter model, the wideband spectrum estimated from the input represents the transfer function of the vocal tract. Both the amplitude and the phase response of the vocal tract are assessed, which is different from most spectral envelope estimators that only consider the amplitude component. However, Psola has a critical flaw: the impulse response of the vocal tract is limited in time to two periods. When the pitch is reduced by less than half, the separation between two impulses becomes longer than the impulse response and a small region in the middle is left equal to zero.

This is where the problem of wideband spectrum interference arises: when neighboring harmonics have different phases, the overlapping spectral content can be attenuated.

This shows that marking the laryngeal pulse instants has a critical effect on the quality of the wideband spectrum, and undesirable peaks can be avoided, to some extent, by centering the window at a local maximum of the absolute value of the signal. Indeed, time domain peaking is a commonly-used time marking method for Psola.

Therefore, a time marking method that minimizes phase interference is needed, in other words, how do you find ΔT in such a way that phase differences for neighboring harmonics are minimized? In most cases, it is impossible to obtain a zero-phase difference, except when the vocal tract filter is zero phase (which is also impossible). Psola thus introduces more or less distortion when the pitch is changed, and the distortion depends on the characteristics of the speaker."

Box 7.1. *The Psola mystery*

7.11.5. *Modification of F0 and duration by phase vocoder*

The phase vocoder (Flanagan and Golden 1965) is an older process in its operating principle, which requires a much greater number of calculations than the Psola method, given that the calculation of two Fourier transforms is necessary, one direct and the other inverse.

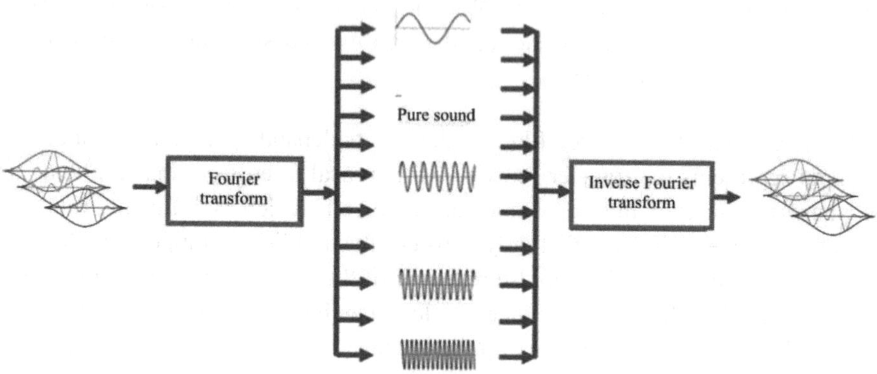

Figure 7.27. *Direct and inverse Fourier analysis of the phase vocoder*

The phase vocoder proceeds by completing a Fourier analysis into a number of harmonic sinusoidal components, the number of which depends on the duration of the time window and the sampling frequency. These components are then processed one by one, either to modify their amplitude, or to extend their temporal validity, so as to modify their duration. The inverse transform of the modified spectrum reconstructs the segment of the sampled signal, and then the successive segments are simply added together to reconstruct the modified signal. Through harmonic analysis, the phase vocoder allows the spectrum to be sculpted between the analysis and additive segment recompositing stages, by modifying the amplitudes of the different harmonics that are provided by the forward transform, before reconstructing speech segments using the inverse transform (Figure 7.27).

The problem with these operations is the phase changes which are introduced by lengthening (obtained by copying segments) or shortening (obtained by deleting segments) the signal durations (which is avoided by the Psola method). If, for example, one wants to increase the speech rate, a certain number of sampled segments are removed, but when the surviving segments are added, their various harmonic components will no longer be in phase and may produce echoes through their addition. The same applies when repeating segments in the signal reconstruction to extend the duration of the signal. It is therefore necessary to correct the phase of each component of each segment in order to achieve a correct and echo-free signal reconstruction, hence the name of the process, phase vocoder (the term *vocoder* comes from *Voice Coding*, used in research on telephonic signal compression).

Figure 7.28 displays an original spectrum, which shows a first modification of F0, then the amplitudes of the harmonics before reconstruction by an inverse Fourier transform.

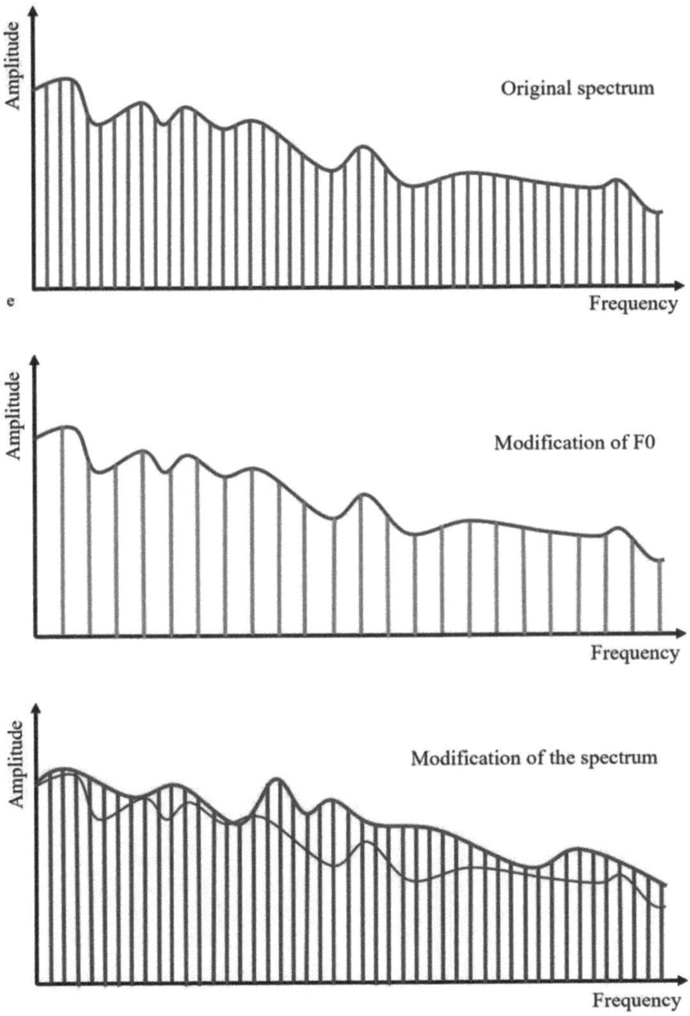

Figure 7.28. *F0 and spectrum modification by phase vocoder*

In film dubbing, a much simpler process is used to adjust the durations of the dubbed speech which are doubled over the duration of the original version. When reproducing the digitized sound, the sample rate is simply accelerated or slowed down. The resulting sound is acceptable for very limited modifications in the order of 5%. Beyond that, distortions that are similar to those of a vinyl disc spinning too fast or too slow are obtained.

8

Articulatory Models

8.1. History

In order to better understand certain phenomena, physicists build models, more or less coherent fabrications simulating the real world in their functioning. In the field of speech production, more or less complex models have been developed, making the best use of available mathematical tools.

For a long time, the principle of explaining the acoustic characteristics of vowels was based on Helmholtz resonators (Hermann von Helmholtz, German physicist, 1821–1894), cavities whose resonant frequency directly depended on their dimensions. These resonators were constructed with different shapes and volumes (pipes, spheres, etc.), whose resonance frequencies corresponded to the various notes of the musical scale. By detecting one of their resonances, an approximate measure of the frequency of a sound could be made (Figure 8.1).

Later, Karl Rudolph Koenig (1832–1901), an ingenious manufacturer of acoustic instruments based in Paris, used this principle and succeeded in making a primitive spectral measurement of speech, thus of a non-stationary sound. The resonance of each sphere was detected by the vibration of the flame of a gas burner, a changing vibration that could be visualized by a system of rotating mirrors. With retinal persistence, it was thus possible to observe the evolution of the spectrum measured (Figure 8.2).

For a color version of all the figures in this chapter, see: www.iste.co.uk/martin/speech.zip.

Figure 8.1. *Helmholtz resonators (source: University of Toronto, Department of Psychology, U of T Scientific Instruments Collection)*

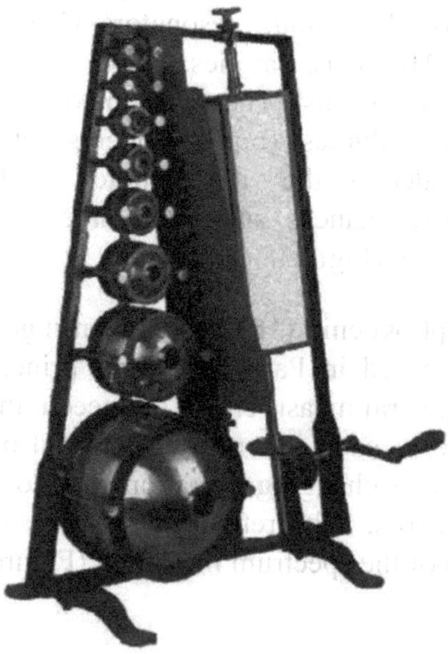

Figure 8.2. *Koening spectral analyzer (source: CNAM 12605)*

From then on, within the community of the first researchers in acoustic phonetics, the idea emerged that resonance, observable by the vocalic formants, is directly related to the volume of the resonant cavity in the vocal tract: a low frequency for the largest cavity and a high frequency for the smallest, depending on the point of articulation dividing the vocal tract. Despite the work on vowels of the precursors Chiba and Kajiyama (1942), and then Fant (1960), it took many years for this doxa to gradually fade away (Martin 2007) in favor of a more accurate conception showing that the frequencies of the formants, and therefore of the resonances, were not each linked to a specific cavity of the vocal tract defined by the articulation of each vowel, but rather of their interaction.

Articulatory models allow a better understanding of the generation of formants, zones of reinforced harmonics. They attempt to mathematically simulate resonance conditions based on a simplified representation of the vocal tract. The necessary approximations are guided by the available mathematical tools, which can only be used for cylindrical volumes or volumes with straight rectangular sections. Thus, sections with a highly variable shape of the vocal tract (Figure 8.3, sections 1 to 10), obtained by casting, scanning or nuclear magnetic resonance on real speakers, will be approximated by circular sections (see for example VocalTractLab, a software implementing a multi-tube model).

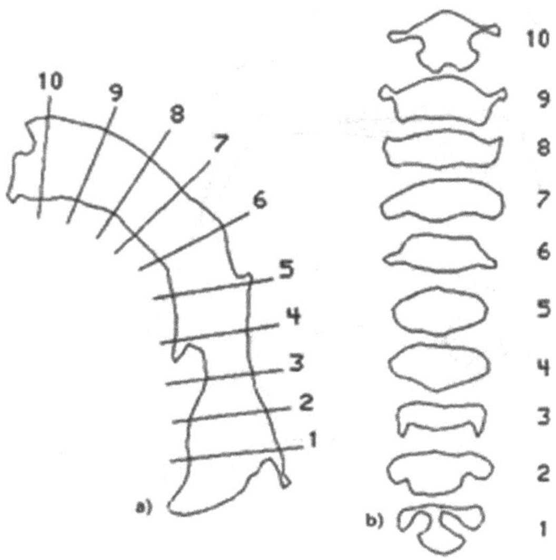

Figure 8.3. *Sections of the vocal tract obtained by molding (Sanchez and Boë 1984)*

Similarly, due to computational constraints, the semi-circular shape of the vocal tract will have to be represented by rectilinear cylinders, with a variable area from section to section. The simplest model only has one cylinder (one tube) and is only suitable for the central vowel [ə]. Despite their relative simplicity, the two-tube models make it possible to account for the frequency distribution of oral vowel formants. Nasal vowels require an additional tube that takes the nasal cavity into account. The n-tube model (for example, with $n = 12$) is the generalization of this technique, which is crucial for understanding the formant frequency distributions of nasal vowels and consonants. It has also made it possible to understand why speech sounds pronounced by young children, who have a smaller vocal tract, could have vocalic formants that are similar enough to those pronounced by adults.

8.2. Single-tube model

The shape of the vocal tract corresponding to the vowel articulation [ə] is the closest to a constant-section tube without sound loss (in reality, the vocal tract is obviously curved, and its section is not really cylindrical) (Figures 8.4 and 8.5).

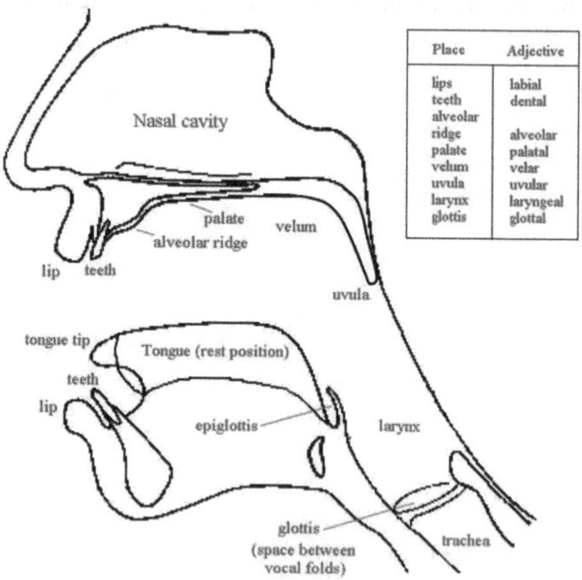

Figure 8.4. *Section showing the articulatory configuration for vowel [ə] (adapted from https://www.uni-due.de/SHE/REV_PhoneticsPhonology.htm)*

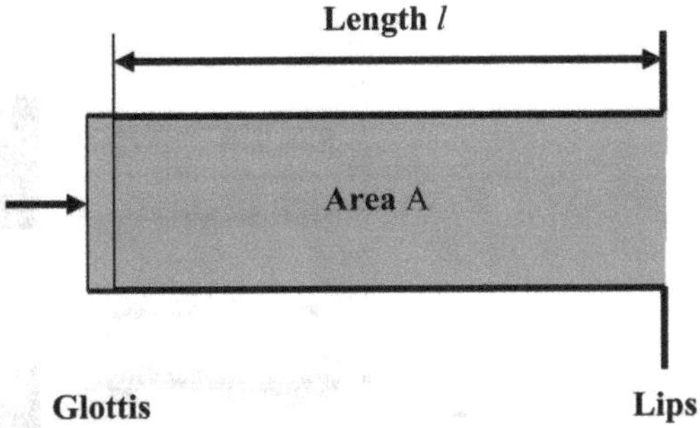

Figure 8.5. *Single-tube vowel model [ə]*

The transfer function of this tube, accounting for the transmission of the harmonics produced by the source (the piston located at the end of the tube) is given by $T(f) = 1/\cos(2\pi fl/c)$, with f = frequency, l = length of the tube and c = speed of sound in (hot) air. We therefore have a resonance for all values of frequency f, which makes the cosine value zero, in other words, when $2\pi fl/c = (2n+1)\pi/2$ with n = 0, 1, 2, ..., n, so for $f = (2n+1)\,c/4l$.

Adopting the values of c = 350 m/s (speed of sound in air at 35°C), and l = 0.175 m as the length of an average male vocal tract, we find a series of resonance values, therefore formants: 500 Hz, 1,500 Hz, 2,500 Hz, 3,500 Hz, etc. There is then not a single formant for this model with a tube corresponding to the articulation of the [ə], but an infinity. This is of course an approximation, since we have neglected the sound losses and damping due to the viscosity of the walls of the vocal tract, the non-cylindrical shape and section of the duct, etc., which are not cylindrical. Moreover, since the source is not impulse-driven, but glottic, with a decrease in harmonic amplitudes in the order of 6 dB to 12 dB per octave, the amplitude and thus the intensity of the harmonics decreases rapidly and is no longer observable in practice below an attenuation of 60 dB to 80 dB.

Figure 8.6 shows the frequency response of the single-tube model. The theoretical formants correspond satisfactorily to those observed on a Fourier harmonic or a Prony spectrum. There is therefore not a single resonant frequency for a single cavity, as has been believed (and written) for a long time in phonetics books, and this frequency does not depend on the volume of the vocal tract for the vowel [ə], but on its length alone.

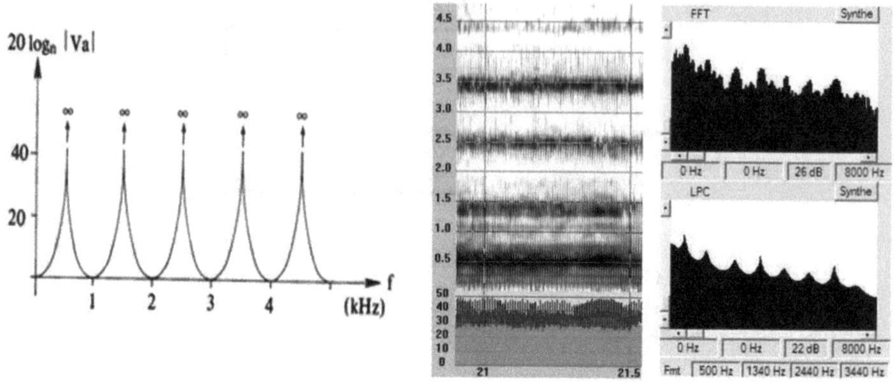

Figure 8.6. *Frequency response for a 17.5 cm 1-tube model, spectrogram, Fourier and Prony spectrum for one vowel [ə]*

8.3. Two-tube model

Traditionally, in articulatory phonetics, oral vowels are described by the rounding or spacing of the lips (which modifies the length of the vocal tract), the opening of the mouth (which modifies the volume of the anterior cavity) and the point of articulation achieved by bringing the back of the tongue closer to the hard palate (which defines a division of the vocal tract into two parts, front cavity, anterior, and back cavity, posterior). A model with two cylindrical tubes and a rectilinear axis is sufficiently adequate to account for this configuration, all the more so since mathematical tools are available to study its properties (Fant 1960).

The two-tube articulatory model is characterized by the following parameters: A_p and l_p area and length of the posterior cavity (upstream of the point of articulation), A_a and l_a area and length of the anterior cavity (downstream of the point of articulation) (Figure 8.7). It can be shown (Fant 1960) that there is resonance for frequency values for which:

$$A_p tg(2\pi f l_a/c) = A_a cotg(2\pi f l_p/c)$$

with c = 350 m/s (speed of sound in air at 35°C), that is to say, whenever a value of frequency f makes the left side equal to the right side. This type of equation, called transcendental (because there is no way to extract the f parameter), can be solved by a graphical method or an algorithmic equivalent to determine the meeting points of the tangent and cotangent trigonometric functions.

For example, to model vowel [a], consider the following values (Figure 8.7):

$$A_\mathrm{p} = 1\ cm^2\ A_\mathrm{a} = 7\ cm^2\ l_\mathrm{p} = 9\ cm\ l_\mathrm{a} = 8\ cm$$

and plot functions $1tg(2\pi f8/c)$ and $7cotg(2\pi f9/c)$ as a function of frequency (Figure 8.8).

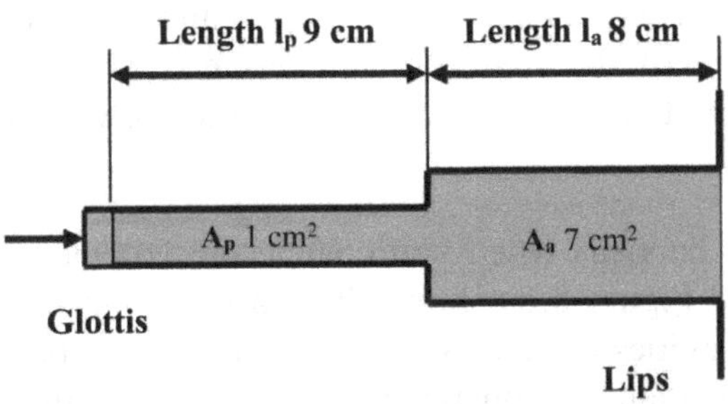

Figure 8.7. *Two-tube model for oral vowels*

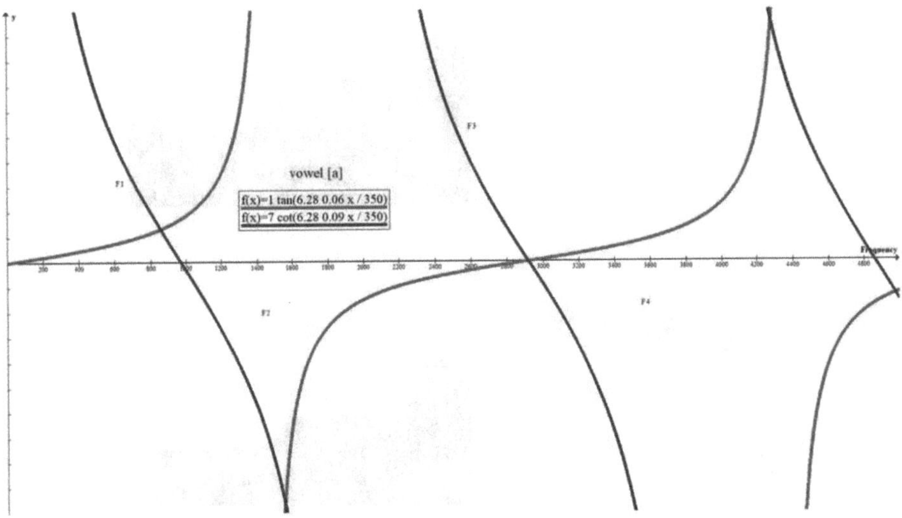

Figure 8.8. *Graphical resolution of the two-tube model of vowel [a] giving the formant frequencies*

The points of intersection correspond to the values of formants:

$$F_1 = 789\ Hz\ F_2 = 1{,}276\ Hz\ F_3 = 2{,}809\ Hz$$

$$F_4 = 3{,}387\ Hz\ F_5 = 4{,}800\ Hz$$

values that compare favorably with the experimental observations in Figure 8.9.

It becomes clear that the frequencies of the formants do not depend directly on each of the anterior and posterior cavities considered in isolation. This has long been written in phonetics literature, where it was explained that the frequency of the first formant was related to the volume of the anterior cavity, and that of the second formant to the volume of the posterior cavity.

What actually happens when you change the volume of the anterior and posterior cavities? Figure 8.10 shows that variations in the areas (and hence volumes) of the cavities do not significantly change the points of intersection of the tangent and cotangent functions, corresponding to the frequencies of the formants.

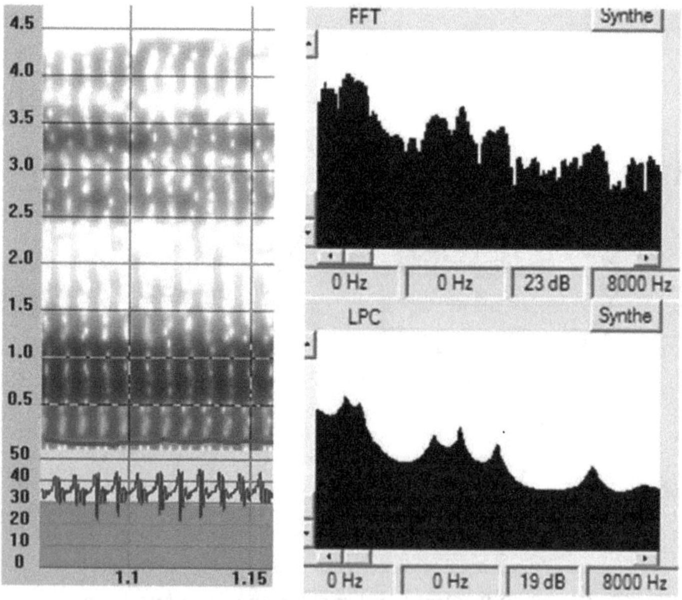

Figure 8.9. *Frequency response for a two-tube model, spectrogram, Fourier and Prony spectrum for vowel [a]*

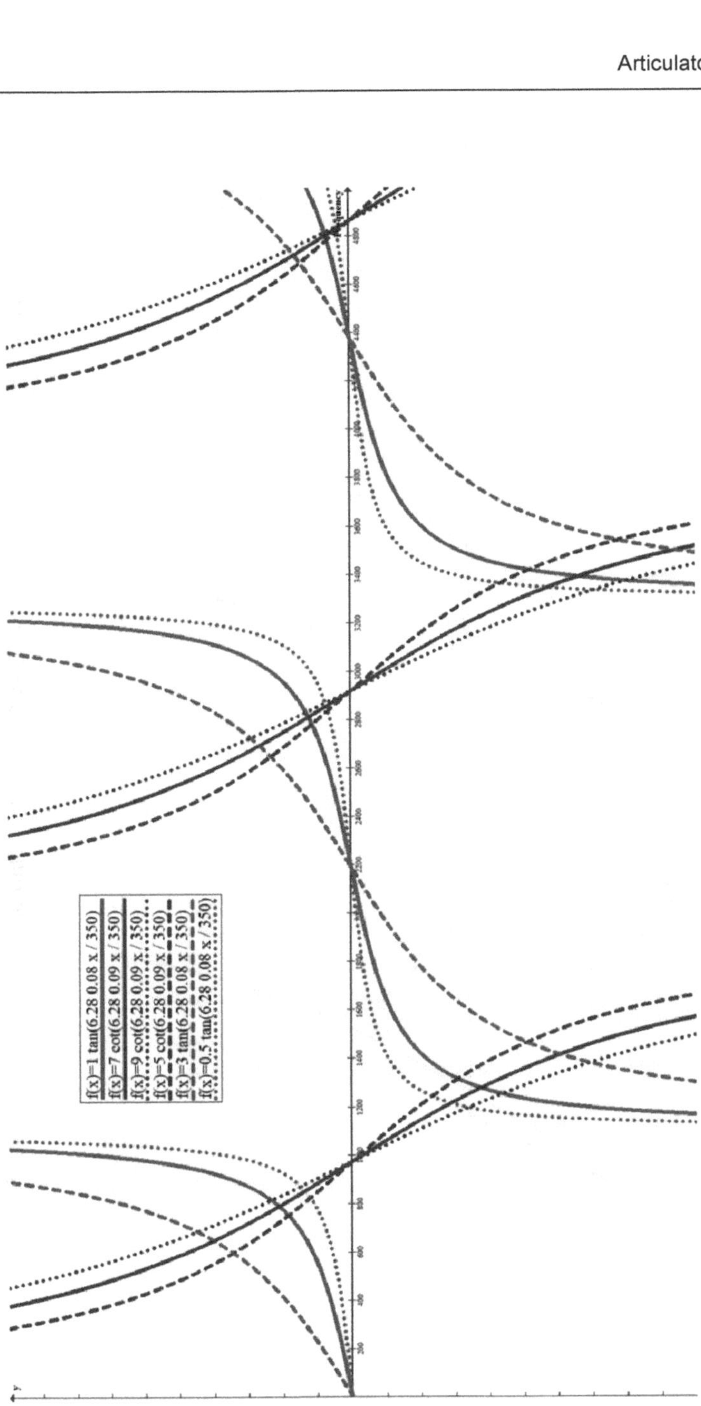

Figure 8.10. *Variations in anterior (5 to 7 cm^2) and posterior (0.5 to 3 cm^2) cavity areas, showing the relative stability of the formant frequencies for a two-tube model*

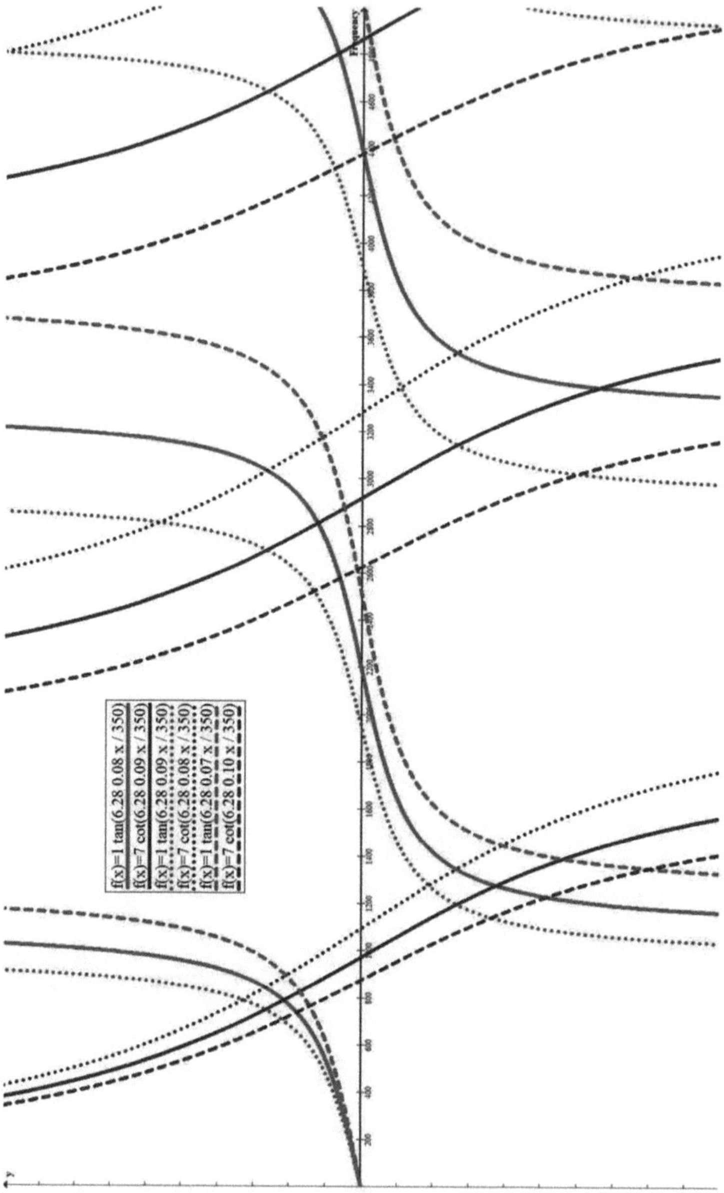

Figure 8.11. *Variations in the ratio of anterior (8 to 10 cm) and posterior (7 to 9 cm) cavity lengths, showing the relative stability of formant frequencies for a two-tube model*

Similarly, the variations in the point of articulation, described in the model by the ratio between the cavity lengths l_a and l_p, lead to few changes in the frequency of formants, as shown in Figure 8.11.

This explains the stability of formant frequencies for designated vowels, allowing relatively important variations of their articulatory configuration.

Figure 8.12 shows some more examples of these configurations.

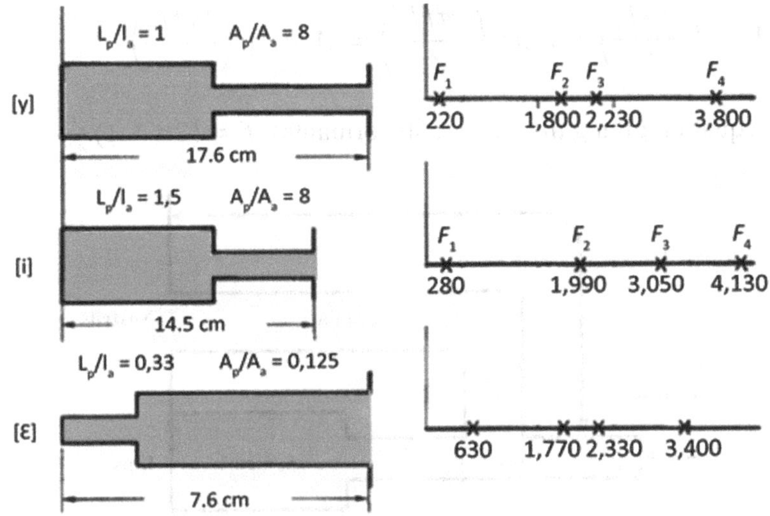

Figure 8.12. *Two-tube models for different oral vowels and corresponding formants*

8.4. Three-tube model

The articulation of a vowel or nasal consonant connects the nasal canal with the vocal tract at about 1/3 of its length from the vocal folds (Figure 8.13). This additional cavity, which is placed in parallel on the two tubes modeling the vocal tract, introduces an additional term in the equation to calculate the formants, as well as a numerator in the transfer function of the whole. Some frequency values will thus give poles (null values of the denominator function of the transfer function), but also zeros (which cancel the numerator of the transfer function). Having the opposite effect of formants, these values are called anti-formants.

Since anti-formants give a zero value for a frequency of the spectrum, it would sometimes be difficult to differentiate them from a simple absence of harmonics at a frequency of the spectrum due to other reasons. It is therefore the understanding of the functioning of the model that allows a satisfactory interpretation to be made of nasal vowel spectra, long described in the literature as being characterized by larger formants.

The equation giving the poles (formants) is:

$$A_a tg\left(\frac{2\pi f l_a}{c}\right) + A_p tg\left(\frac{2\pi f l_p}{c}\right) = A_n cotg\left(\frac{2\pi f l_n}{c}\right)$$

The equation giving the zeros (anti-formants): $f = (2n+1)\frac{c}{4l_a}$.

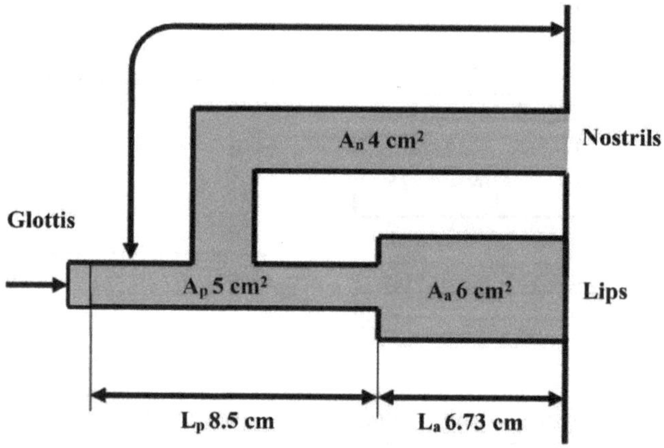

Figure 8.13. *Model with three tubes of [m] (based on Flanagan (1965))*

We therefore have an anti-formant for A_1 = 1,300 Hz, A_2 = 3,900 Hz, A_3 = 6,500 Hz, etc.

The formant frequencies correspond to the intersection of the cotangent with the sum of the two tangents of the equation: F_1 = 250 Hz, F_2 = 1,150 Hz, F_3 = 1,350 Hz, F_4 = 2,200 Hz. Formants (poles) and anti-formants (zeros) are often represented on a frequency axis by a cross (x) and a zero (o), as shown in Figure 8.14 for the nasal consonant [m].

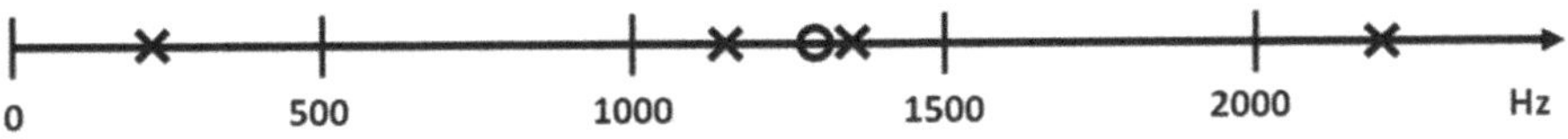

Figure 8.14. *Distribution of formants and anti-formants for the nasal consonant [m]*

This example shows that the first zero at 1,300 Hz is placed between two poles that are very close in frequency, 1,150 Hz and 1,350 Hz (Figure 8.15). Observed on a spectrogram, the second and third formants will appear muddled due to the insufficient frequency resolution of the Fourier harmonic analysis, and the anti-formant therefore cannot be detected. This accounts for early measurements of nasal vowel formants with a second formant that is larger than the second formant of the corresponding oral vowels (Figure 8.16).

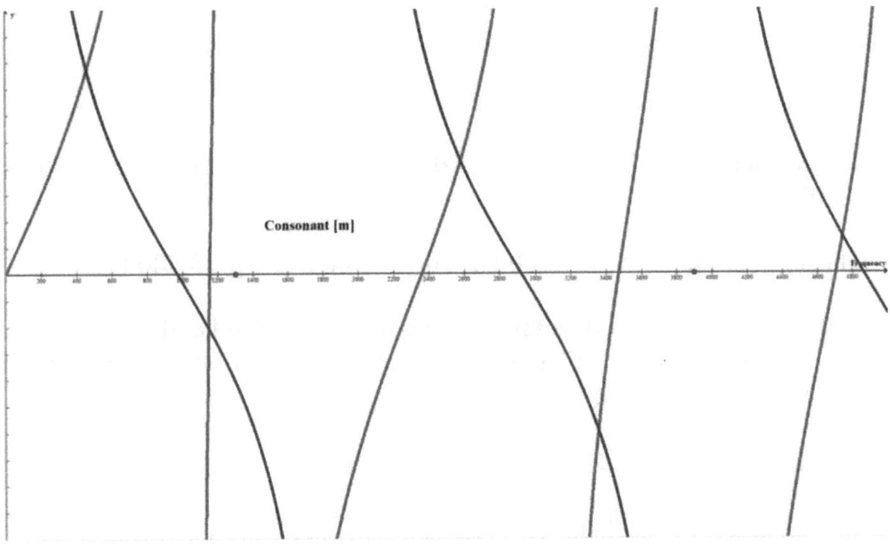

Figure 8.15. *Graphic resolution of the three-tube model of the nasal consonant [m]. The red dots represent the anti-formants*

Another example pertains to the nasal vowel [ã], with $A_a = 7\ cm^2$, $l_a = 8$ cm, $A_p = 1$ cm, $l_p = 9$ cm, $A_n = 4\ cm^2$, $l_n = 12.5$ cm, corresponding to the graph in Figure 8.16.

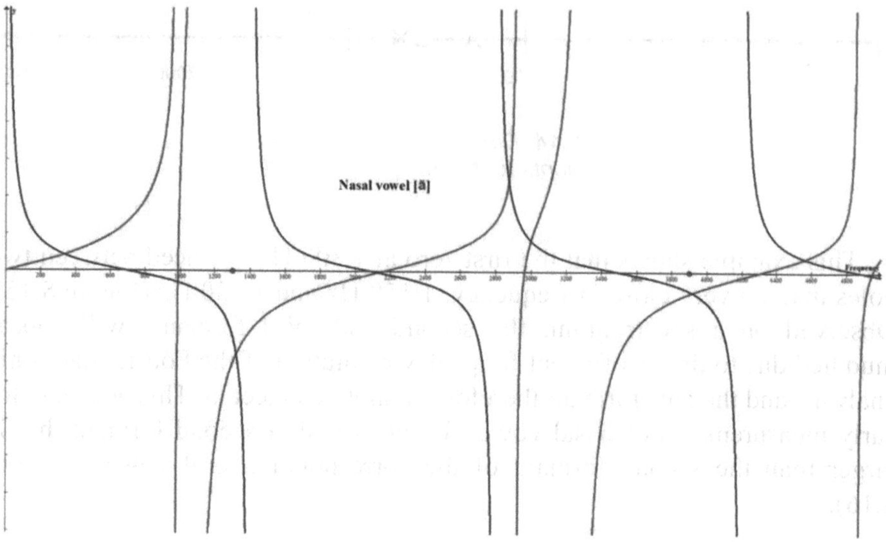

Figure 8.16. *Graphic resolution of the three-tube nasal vowel model [ã]. The red dots represent the anti-formants*

The formants: F_1 = 350 Hz, F_2 = 1,000 Hz, F_3 = 1,250 Hz, F_4 = 2,150 Hz, F_5 = 3,000 Hz.

The anti-formants: A_1 = 1,100 Hz, A_2 = 3,300 Hz, A_3 = 5,500 Hz.

The first anti-formant thus appears between the second and third formant, giving the impression of a second formant larger than that of the corresponding oral vowel.

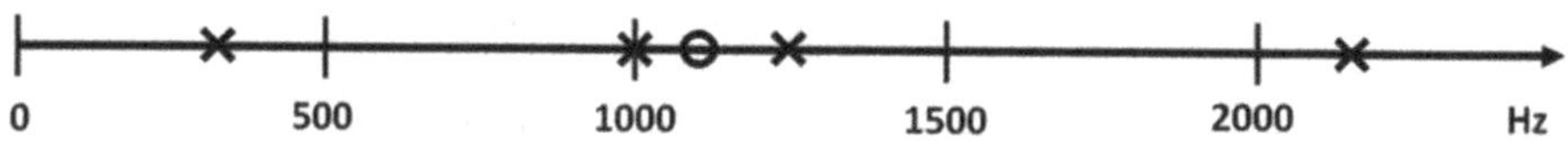

Figure 8.17. *Distribution of formants and anti-formants for the nasal vowel [ã]*

Figures 8.18a and 8.18b show experimental observations of vowels [a] (oral) and [ã] (nasal), whose formant frequencies compare nicely with values predicted by the model.

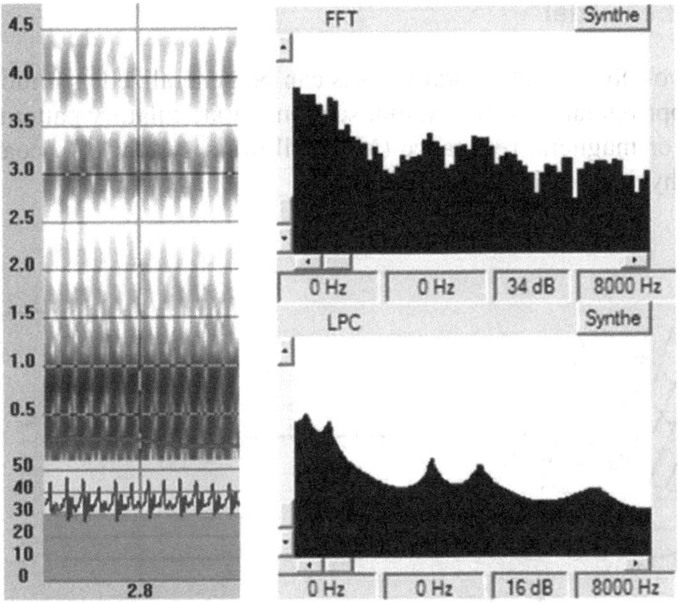

Figure 8.18a. *Fourier and Prony spectrogram and spectra, nasal vowel [ã] above, oral vowel [a] below*

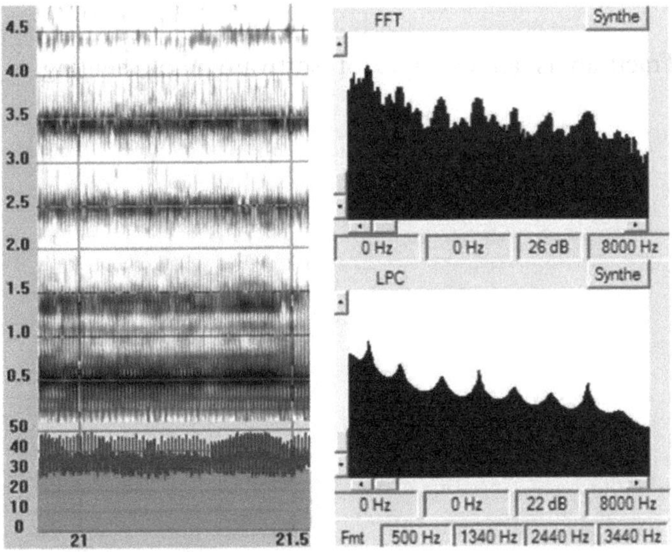

Figure 8.18b. *Fourier and Prony spectrogram and spectra, nasal vowel [ã] above, oral vowel [a] below (continued)*

8.5. N-tube model

The two-tube model of oral vowels can be generalized as n tubes (Maeda 1979). Approximation from sagittal sections of articulatory patterns obtained by X-ray or magnetic resonance (MRI) allows a system of n coaxial tubes, close to physical reality, to be defined.

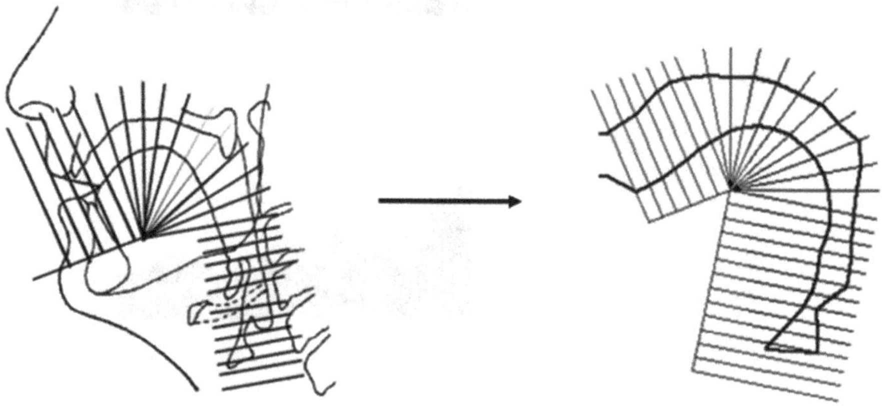

Figure 8.19. *Model with n tubes obtained by segmentation of sagittal sections (Maeda 1979)*

VocalTractLab is an example of software implementing a multi-tube model.

Appendix

A.1. Definition of the sine, cosine, tangent and cotangent of an angle α

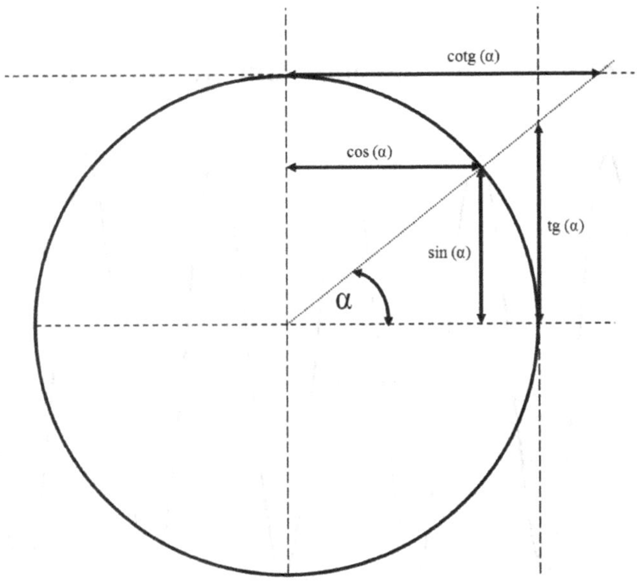

Figure A.1. *Trigonometric circle*

To define the sine, cosine, tangent and cotangent trigonometric functions, we refer to a circle of unit radius (in other words, a circle with a radius of 1), and to two perpendicular axes passing through the center of the circle: one horizontal, the other vertical. A straight line is then defined starting from the center of the circle and forming an angle α with the horizontal axis of the circle.

The sine of angle α is then equal to the distance from the horizontal axis to this line's point of intersection with the circle.

The cosine of angle α is equal to the distance from the vertical axis to this line's point of intersection with the circle.

The tangent is equal to the distance from this line's point of intersection with another line that is perpendicular to the horizontal axis at the point of intersection of the horizontal axis and the circle.

The cotangent is equal to the distance from this line's point of intersection with another line that is perpendicular to the vertical axis at the point of intersection of the vertical axis and the circle.

A.2. Variations of sine, cosine, tangent and cotangent as a function of angle α

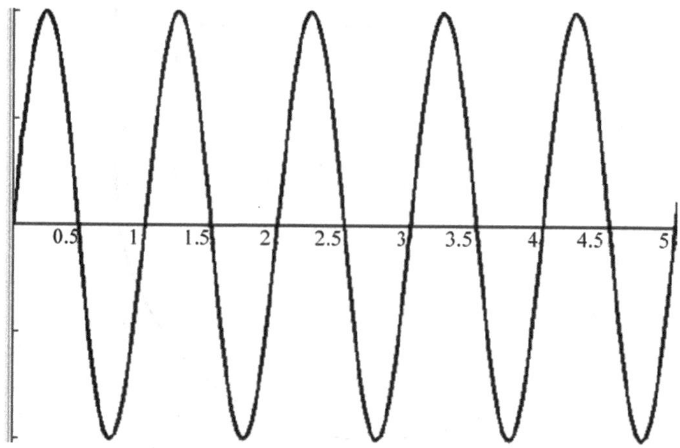

Figure A.2. *Sine function sin (α)*

The value of the sine starts from zero with angle α equal to zero. It reaches 1 when α is 90 degrees (in other words, π/2), then zero again when α = 180 degrees (π), −1 when α = 270 degrees (3π/2), and finally zero after one complete revolution, in other words, when α = 360 degrees (2π). The cycle then starts again.

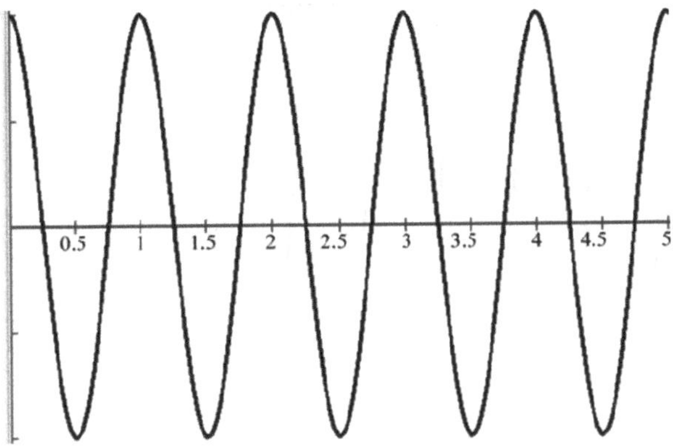

Figure A.3. *Cosine function cos (α)*

The value of the cosine starts from 1 with angle α equal to zero. It reaches 0 when α is 90 degrees (in other words, π/2), then −1 when α = 180 degrees (π), 0 when α = 270 degrees (3π/2), and finally 1 again after one complete revolution, in other words, when α = 360 degrees (2π).

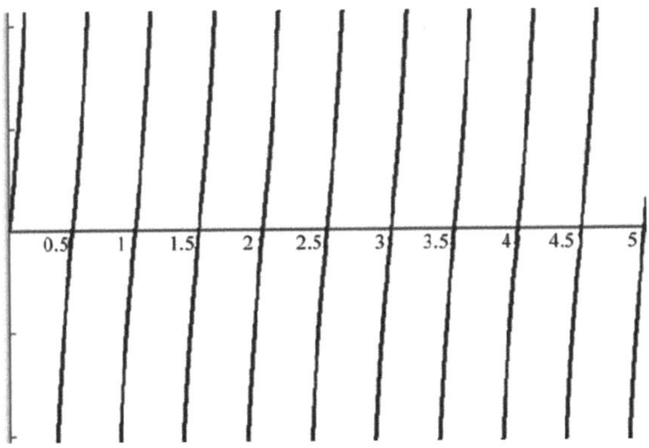

Figure A.4. *Tangent function tg (α)*

The value of the tangent starts from zero with angle α equal to zero. It reaches 1 when α is 45 degrees (in other words, π/4), then infinity (∞) for 90 degrees (π/2). Then, it changes abruptly to negative infinity (−∞) to go back to −1 when α = 135 degrees (3π/4), then again to zero when

α = 180 degrees (π), $-\infty$ negative infinity when α = 270 degrees ($3\pi/2$), and finally zero after a complete revolution, that is to say, when α = 360 degrees (2π). The cycle then starts again. The tangent is therefore always increasing.

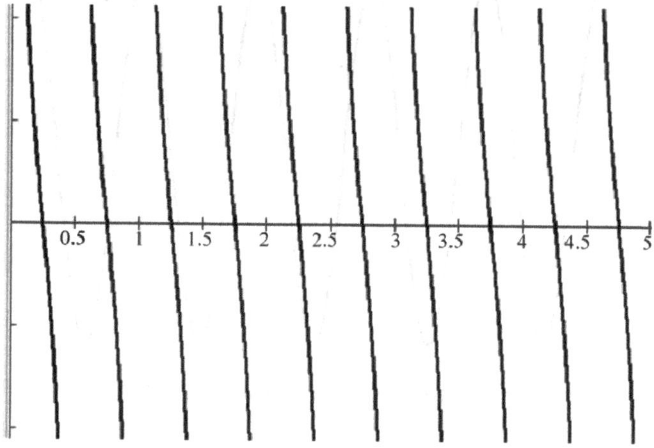

Figure A.5. *Cotangent function cotg (α)*

The value of the cotangent starts from ∞ with angle α equal to zero. It goes down to zero for 90 degrees ($\pi/2$). It then continues to descend to negative infinity ($-\infty$), suddenly changes to positive infinity (∞) when α is 180 degrees (π), descends to zero when α = 270 degrees ($3\pi/2$), and finally $-\infty$ after a complete revolution, that is to say, when α = 360 degrees (2π). The cotangent is therefore always decreasing.

References

de Abromont, C. and Montalembert, E. (2001). *Le Guide de la théorie de la musique*. Fayard/Henry Lemoine, Paris.

Avanzi, M. (2013). Note de recherche sur l'accentuation et le phrasé à la lumière des corpus du français. *Tranel*, 58, 5–24.

Bailly, L., Henrich, N., Pelorso, X. (2010). Vocal fold and ventricular fold vibration in period-doubling phonation: Physiological description and aerodynamic modeling. *JASA*, 5(127), 3212–3222.

Beckman, M.E., Hirschberg, J., Shattuck-Hufnagel, S. (2005). The original ToBI system and the evolution of the ToBI framework. In *Prosodic Typology – The Phonology of Intonation and Phrasing*, S.-A. Jun (ed.). Oxford University Press, Oxford.

Boë, L.-J. (2000). Forensic voice identification in France. *Speech Communication*, 31(2/3), 205–224.

Boë, L.-J. and Rakotofiringa, H. (1971). Exigences, réalisation et limites d'un appareillage destiné à l'étude de l'intensité et de la hauteur d'un signal acoustique. *Revue d'acoustique*, 4, 104–113.

Calliope (1989). *La Parole et son traitement automatique*. Masson, Paris.

Camacho, A. (2007). Swipe: A sawtooth waveform inspired pitch estimator for speech and music. PhD thesis, University of Florida, Gainesville.

Carré, R. (2004). From an acoustic tube to speech production. *Speech Communication*, 42(2), 227–240.

Chen, C.J. (2016). *Elements of Human Voice*. Word Scientific Publishing Co., Singapore.

Chiba, T. and Masato, K. (1942). *The Vowel. Its Nature and Structure*. Kaiseikan, Tokyo.

Christodoulides, G. and Avanzi, M. (2014). An evaluation of machine learning methods for prominence detection in French. In *Proc. Interspeech 2014*. Singapore, 116–119.

Cooley, J.W. and Tukey, J.W. (1965). An algorithm for the machine calculation of complex Fourier series. *Mathematical Computing*, 19, 297–301.

Delais-Roussarie, E., Post, B., Avanzi, M., Buthka, C., Di Cristo, A., Feldhausen, I., Jun, S.-A., Martin, P., Meisenburg, T., Rialland, A., Sichel-Bazin, R., Yoo, H.-Y. (2015). Intonational phonology of French: Developing a ToBI system for French. In *Intonational in Romance*, Frota, S., Prieto, P. (eds). Oxford University Press, Oxford.

Fant, G. (1960). *Acoustic Theory of Speech Production*. Mouton, La Haye.

Flanagan, J.L. (1965). *Speech Analysis: Synthesis and Perception*. Springer, Heidelberg.

Flanagan, J.L. and Golden, R.M. (1965). Phase vocoder. *Bell System Technical Journal*, 45, 1493–1509.

Fletcher, H. and Munson, W.A. (1933). Loudness, its definition, measurement and calculation. *Journal of the Acoustical Society of America*, 5, October, 82–108.

Fourier, J.-B.J. (1822). *Théorie analytique de la chaleur*. Firmin-Didot, Paris.

Ghasemzadeh, N. and Zafari, A.M. (2011). A brief journey into the history of the arterial pulse. *Cardiology Research and Practice*, 1.

Goldman, J.-P. (2020). EasyAlign: Phonetic alignment with Praat [Online]. Available at: http://latlcui.unige.ch/phonetique/easyalign.php.

Haas, H. (1972). The influence of a single echo on the audibility of speech. *Audio Engineering Society*, March, 145–159.

Haynes, B. (2002). *A History of Preforming Pitch. The Story of "A"*. Scarecrow, Lanham.

Henrich, N. (2001). Étude de la source glottique en voix parlée et chantée : modélisation et estimation, mesures acoustiques et électroglottographiques, perception. PhD thesis, Université Paris VI, Paris.

Henrich, N., D'Alessandro, C., Castellengo, M., Doval, B. (2005). Glottal open quotient in singing: Measurements and correlation with laryngeal mechanisms, vocal intensity, and fundamental frequency. *Journal of the Acoustical Society of America*, 117(3), 1417–1430.

Hess, W. (1983). *Pitch Determination of Speech Signals*. Springer Verlag, New York.

Hollien, H. and Michel, J. (1968). Vocal fry as a phonational register. *J. Speech Hear Res.*, 11(3), September, 600–604.

Hollien, H., Michel, J., Doherty, T.E. (1973). A method for analyzing vocal jitter in sustained phonation. *Journal of Phonetics*, 1, 8591.

Hua, K. (2015). Reunderstand PSOLA [Online]. Available at: https://www.academia.edu/29945658/Reunderstand_PSOLA.

Juszkiewicz, R. (2020). Speech production using concatenated tubes, EEN 540 – Computer project II. *Qualitybyritch* [Online]. Available at: http://www.qualitybyrich.com/een540proj2/.

Léon, P.R. and Martin, P. (1970). *Prolégomènes à l'étude des structures intonatives.* Didier, Montreal.

Maeda, S. (1979). Un modèle articulatoire de la langue avec des composantes linéaires. In *Actes des 10ème journées d'étude sur la parole*. Grenoble, May, 152–162.

Malfrère, F. and Dutoit, T. (1997). High-quality speech synthesis for phonetic speech segmentation. In *Proc. Eurospeech 1997*, Kokkinakis, G., Fakotakis, N., Dermatas, E. (eds). Rhodes.

Mallat, S. (1989). A theory for multiresolution signal decomposition: The wavelet representation. *IEEE Transaction on Pattern Analysis and Machine Intelligence*, 11, July, 674–693.

Martin, P. (1982). Comparison of pitch detection by cepstrum and spectral comb analysis. In *Proceedings of the 1982 IEEE International Conference on Acoustics, Speech, and Signal Processing*, Gueguen, C. (ed.). Paris.

Martin, P. (2000). Peigne et brosse pour F0 : mesure de la fréquence fondamentale par alignement de spectres séquentiels. *Actes des 23ème journées d'étude sur la parole*, 245–248.

Martin, P. (2007). Les formants vocaliques et le barrissement de l'éléphant. *Histoire des théories linguistiques*, X, 9–27.

Martin, P. (2018a). Un algorithme de segmentation en phrasé. *Actes des 32ème journées d'étude sur la parole* [Online]. Available at: https://www.isca-speech.org/archive/JEP_2018/.

Martin, P. (2018b). *Intonation, structure prosodique et ondes cérébrales*. ISTE Editions, London.

McKinney, N.P. (1965). Laryngeal frequency analysis for linguistic research. Report, University of Michigan, Ann Arbor.

Michel, U. (2016). 432 contre 440 Hz, l'étonnante histoire de la guerre des fréquences, 17 September [Online]. Available at: http://www.slate.fr/story/118605/frequences-musique.

Moulines, É., Charpentier, F., Hamon, C. (1989). A diphone synthesis system based on time-domain prosodic modifications of speech. In *Proceedings of the 1989 IEEE International Conference on Acoustics, Speech, and Signal Processing*, Gueguen, C. (ed.). Paris.

ORFEO (2020). Corpus d'étude pour le français contemporain [Online]. Available at: www.projet-orfeo.fr.

PRAAT (2020). Doing phonetic with computers [Online]. Available at: www.praat.org.

de Prony, G.R. (1795). Essai expérimental et analytique : sur les lois de la dilatabilité de fluides élastiques et sur celles de la force expansive de la vapeur de l'alkool, à différentes températures. *Journal de l'École polytechnique*, 1(22), 24–76.

Robinson, D.W. and Dadson, R.S. (1956). Plots of equal loudness as a function of frequency. *British Journal of Applied Physics*, 7, 166.

Rossi, M. (1971). Le seuil de glissando ou seuil de perception des variations tonales pour la parole. *Phonetica*, 23, 1–33.

Sanchez, H. and Boë, L.-J. (1984). De la coupe sagittale à la fonction d'aire du conduit vocal. *Actes des 13ème journées d'étude sur la parole*. Brussels, 23–25.

Stevens, S.S. (1957). On the psychophysical law. *Psychological Review*, 64(3), 153–181.

Sundberg, J. (1977). The acoustics of the singing voice. *Scientific American*, 236, 3.

Taylor, P. (2009). *Text-to-Speech Synthesis*. Cambridge University Press, Cambridge [Online]. Available at: http://research.cs.tamu.edu/prism/lectures/sp/l9.pdf.

Teston, B. (2006). À la poursuite du signal de parole. *Actes des 26ème journées d'étude sur la parole*. Aussois, June, 7–10.

Vaissière, J. (2006). *La phonétique*. PUF, Paris.

Yamagishi, J., Honnet, P.-É., Garner, P., Lazaridis, A. (2016). The SIWIS French speech synthesis database. University of Edinburgh, Edinburgh [Online]. Available at: https://doi.org/10.7488/ds/1705.

Yu, K.M. (2010). Laryngealization and features for Chinese tonal recognition. *Proc. Interspeech-2010*. Chiba, 1529–1532.

WINPITCH (N/A). WinPitch, speech analysis software [Online]. Available at: www.winpitch.com.

Index

M, N

O, P

Q, R, S

T, V

Other titles from

iSTE

in

Cognitive Science and Knowledge Management

2020

GILLIOZ Christelle, ZUFFEREY Sandrine
Introduction to Experimental Linguistics

NGUYEN-XUAN Anh
Cognitive Mechanisms of Learning
(Learning, Development and Cognitive Technologies Set – Volume 1)

OSIURAK François
The Tool Instinct

ZUFFEREY Sandrine
Introduction to Corpus Linguistics

2019

CLAVEL Chloé
Opinion Analysis in Interactions: From Data Mining to Human-Agent Interaction

KAROUI Jihen, BENAMARA Farah, MORICEAU Véronique
Automatic Detection of Irony: Opinion Mining in Microblogs and Social Media

MARTINOT Claire, BOŠNJAK BOTICA Tomislava, GEROLIMICH Sonia, PAPROCKA-PIOTROWSKA Urszula
Reformulation and Acquisition of Linguistic Complexity: Crosslinguistic Perspective
(Interaction of Syntax and Semantics in Discourse Set – Volume 2)

2018

BONFANTE Guillaume, GUILLAUME Bruno, PERRIER Guy
Application of Graph Rewriting to Natural Language Processing
(Logic, Linguistics and Computer Science Set – Volume 1)

PENNEC Blandine
Discourse Readjustment(s) in Contemporary English
(Interaction of Syntax and Semantics in Discourse Set – Volume 1)

2017

KURDI Mohamed Zakaria
Natural Language Processing and Computational Linguistics 2: Semantics, Discourse and Applications

MAESSCHALCK Marc
Reflexive Governance for Research and Innovative Knowledge
(Responsible Research and Innovation Set – Volume 6)

PELLÉ Sophie
Business, Innovation and Responsibility
(Responsible Research and Innovation Set - Volume 7)

2016

BOUVARD Patricia, SUZANNE Hervé
Collective Intelligence Development in Business

CLERC Maureen, BOUGRAIN Laurent, LOTTE Fabien
Brain–Computer Interfaces 1: Foundations and Methods
Brain–Computer Interfaces 2: Technology and Applications

FORT Karën
Collaborative Annotation for Reliable Natural Language Processing

GIANNI Robert
Responsibility and Freedom
(Responsible Research and Innovation Set – Volume 2)

GRUNWALD Armin
The Hermeneutic Side of Responsible Research and Innovation
(Responsible Research and Innovation Set – Volume 5)

KURDI Mohamed Zakaria
Natural Language Processing and Computational Linguistics 1: Speech, Morphology and Syntax

LENOIR Virgil Cristian
Ethical Efficiency: Responsibility and Contingency
(Responsible Research and Innovation Set – Volume 1)

MATTA Nada, ATIFI Hassan, DUCELLIER Guillaume
Daily Knowledge Valuation in Organizations

NOUVEL Damien, EHRMANN Maud, ROSSET Sophie
Named Entities for Computational Linguistics

PELLÉ Sophie, REBER Bernard
From Ethical Review to Responsible Research and Innovation
(Responsible Research and Innovation Set - Volume 3)

REBER Bernard
Precautionary Principle, Pluralism and Deliberation
(Responsible Research and Innovation Set – Volume 4)

SILBERZTEIN Max
Formalizing Natural Languages: The NooJ Approach

2015

LAFOURCADE Mathieu, JOUBERT Alain, LE BRUN Nathalie
Games with a Purpose (GWAPs)

SAAD Inès, ROSENTHAL-SABROUX Camille, GARGOURI Faïez
Information Systems for Knowledge Management

2014

DELPECH Estelle Maryline
Comparable Corpora and Computer-assisted Translation

FARINAS DEL CERRO Luis, INOUE Katsumi
Logical Modeling of Biological Systems

MACHADO Carolina, DAVIM J. Paulo
Transfer and Management of Knowledge

TORRES-MORENO Juan-Manuel
Automatic Text Summarization

2013

TURENNE Nicolas
Knowledge Needs and Information Extraction: Towards an Artificial Consciousness

ZARATÉ Pascale
Tools for Collaborative Decision-Making

2011

DAVID Amos
Competitive Intelligence and Decision Problems

LÉVY Pierre
The Semantic Sphere: Computation, Cognition and Information Economy

LIGOZAT Gérard
Qualitative Spatial and Temporal Reasoning

PELACHAUD Catherine
Emotion-oriented Systems

QUONIAM Luc
Competitive Intelligence 2.0: Organization, Innovation and Territory

2010

ALBALATE Amparo, MINKER Wolfgang
Semi-Supervised and Unsupervised Machine Learning: Novel Strategies

BROSSAUD Claire, REBER Bernard
Digital Cognitive Technologies

2009

BOUYSSOU Denis, DUBOIS Didier, PIRLOT Marc, PRADE Henri
Decision-making Process

MARCHAL Alain
From Speech Physiology to Linguistic Phonetics

PRALET Cédric, SCHIEX Thomas, VERFAILLIE Gérard
Sequential Decision-Making Problems / Representation and Solution

SZÜCS Andras, TAIT Alan, VIDAL Martine, BERNATH Ulrich
Distance and E-learning in Transition

2008

MARIANI Joseph
Spoken Language Processing

Printed in the USA/Agawam, MA
September 16, 2026

910661.020